MW01115308

ELECTRONIC CONTROL PROJECTS FOR THE HOBBYIST AND TECHNICIAN

by Henry C. Smith and Craig B. Foster

PROMPT Publications® is an imprint of Howard W. Sams & Company, 2647 Waterfront Parkway, East Drive, Indianapolis, IN 46214-2041.

International Standard Book Number: 0-7906-1044-2

Editor: Candace M. Drake

Assistant Editor: Rebecca A. Hartford

Editorial Assistant: Kelly A. Bell

Art Work: Kenneth Cobb, Joseph Kocha

Cover Design: Sara Wright

Typesetting: Leah Marckel

Acknowledgements

All photographs not credited are either courtesy of Author or Howard W. Sams & Company.

Printed in the United States of America

9 8 7 6 5 4 3 2 1

TABLE OF CONTENTS

INTRODUCTION

If you are interested in learning more about electronics, one of the most enjoyable ways of gaining more knowledge and practical experience is by building projects. Taking the time to examine a circuit so that you understand how and why it works, and then actually building it, helps in both areas. It is here that you learn how all the theory comes together to accomplish useful things. We are assuming, of course, that you already know basic electronic theory. If not, you may wish to select a good book on this subject to use in conjunction with this one. This book was not written as a textbook to expound electronics theory in a systematic manner, but to demonstrate how a knowledge of the theory can be used to synthesize circuits which will solve real problems.

To make the projects in this book as educational as possible, we deliberately avoided a step-by-step approach; you will have to do some thinking for yourself. We do, however, provide a complete schematic for each circuit, along with a detailed description of how it works, what function the various components perform, and troubleshooting guidelines. Where appropriate, we also point out how the circuit could be modified to perform a different task or function. Our purpose throughout has been not only to give you a useful project that will work, but to involve you as much as possible in the synthesis of the circuit. When you finish a project, we want you not only to know how to build it, but also how and why it works as a complete unit. We try to avoid the common practice of throwing circuit parts at you; there are no 40 kHz Oscillator or Squarewave Oscillator projects here, although these items are found as components in some of the circuits presented.

Judging from our own experiences, we know that nothing is more discouraging to the electronics hobbyist than to spend time and money building a project which will not work when completed. Yet, it is not at all unusual to pick up an electronics project book and, upon close examination, discover that half of the projects could not possibly work. In some cases, the circuit may actually work for a time, but is doomed to an early failure due to poor design. We have tried our best to give you reliable circuits. We have built and used every one to make certain that they work and are dependable. In some cases, you may have to experiment for optimum results. In the plant watering project, for example, you may find that you will have to experiment with probe placement. Our geographical area has a very heavy type of clay soil; we honestly do not know what effect, if any, other types of soil will have on sensor placement. We can guarantee, however, that the circuit has been tested and will work.

Many different construction techniques can be used to build an electronic circuit, but the only one we used in these projects (and the only one we really recommend) is construction on a printed circuit board. This type of construction is hard to beat for sturdiness, reliability, and permanence. Although you can always make your own printed circuit boards, for relatively simple circuits the easiest method is to use pre-etched boards that are available at Radio Shack and other electronics stores. These boards come with various patterns of copper foil already etched and can be a real time-saver. We used this type of board for all of the projects in this book. One helpful hint: before beginning construction, always clean the copper surface thoroughly with an abrasive pad. If you do not, the solder will not stick to the copper.

If you have never built any power line-operated projects before, we urge you to be extremely careful. The power line voltage can be lethal. Do not take any chances that might allow a person to come into contact with line voltage. Always make sure that the DC portion of the circuit you are building is completely isolated from the AC line. If necessary, find someone more experienced than yourself to help you through the first few projects. We wish we could avoid this hazard altogether and simply run these circuits on batteries. Many project books take this approach, or avoid the problem altogether by simply showing an arrow going to +12 v. The truth of the matter is that most projects like these must be line-operated in order to be practical. Most of the circuits in this book are geared to 'control' applications and must be left running continuously to be useful. Unless a circuit will run reliably on a battery for a year or more, this approach is too much of a nuisance.

As stated previously, we assume that you are already familiar with basic electronics theory and have some knowledge of the function of the more common components, such as resistors, capacitors, diodes, and transistors. We decided to give you a brief description of the integrated circuits used in these projects, just in case you are not familiar with them. Below, we have divided all of the ICs used in this book into one of two categories: analog and digital. The analog section includes ICs like the op-amp, linear timers, and other chips which deal with varying analog voltages. The digital section includes logic gates and flip-flops--ICs whose primary function is to deal with digital quantities. Some of these ICs are covered extensively in a particular project, so they are mentioned only briefly here. Also, the SSI202P Tone Decoder used in the last project will not be addressed, because it is a very complicated chip whose operation can best be understood by studying its data sheet, which is several pages long.

ANALOG INTEGRATED CIRCUITS

The Op-Amp

The operational amplifier, or op-amp, is a high gain amplifier with a differential input. Instead of having a single input referenced to circuit ground, the op-amp has two inputs referenced to each other. One input is known as the inverting input (indicated by a "-" sign), and the other is known as the non-inverting input (indicated by a "+" sign). The schematic symbol for an op-amp is shown in figure 1.

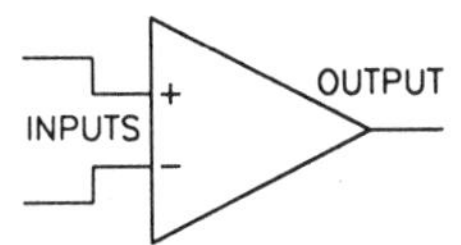

Figure 1 The schematic symbol for an op-amp.

The output of the op-amp is dependent upon the relationship between the two inputs. If the voltages on the two inputs are exactly equal, the output should be one-half of the supply voltage on an ideal op-amp. Often, a dual power supply is used with a plus and minus output, so that the output will be 0 v when the inputs match. If the non-inverting (+) input is made slightly more positive than the inverting input, the output will start to go more positive. If the non-inverting input is made slightly more negative than the inverting input, the output will swing in the negative direction. Note that the gain of the op-amp is very high, often 200,000 or more. As a result, only a small difference between the two inputs will cause the output to swing as far as the power supply voltages will allow in either direction. Usually, on an op-amp with a bipolar output (such as the 741 and 1458 used in these projects), the output will only rise to within a volt or two of the maximum supply voltage and will only drop to within a volt or two of the negative side of the supply. CMOS op-amps are also available, and their outputs will usually swing very close to either supply rail.

Because the gain of these units is so high, some negative feedback is usually provided to limit the gain to some fixed amount or to cause the circuit to behave in some specific way. Occasionally, however, it is desirable to use this high gain without any limiting and run the op-amp "wide open." This design is usually used when it is necessary to detect whether a certain level or threshold has been exceeded. An op-amp designed to be used in this way is known as a "comparator." The op-amps in this book have all been used in the wide-open mode.

The 555 Linear Timer

Several of the projects in this book make use of the 555 linear timer IC. This versatile device can be used in either an astable mode (as a signal source or oscillator) or in a monostable mode (as a timer or one-shot). We will not elaborate on the internal structure and hook-up of the 555 here, since they are described sufficiently elsewhere in this book. For a description of its internal structure, see the 'Light Sequencer' project. Its use in both astable and monostable modes is detailed in the 'Long-term Linear Timer' project.

The Linear Regulator

The linear regulator IC provides a simple, economical method of regulating the output of a power supply. Without a regulator of some type, the output of a typical capacitively-filtered supply will drop as the load is increased. This is acceptable for some simple circuits, but many ICs require a power supply voltage to be held within a very tight range. For instance, most TTL logic ICs require a supply of 5 v, held within plus or minus 1/4 v. Many other ICs do not require such tight regulation, but circuit performance and reliability are enhanced by holding the supply steady. This is especially true of circuits where very small voltages are being compared, or circuits where a relatively heavy load, such as a relay, is being switched on and off. When you are measuring voltages in the mV range, you do not need the power supply voltage going up and down by a volt or more.

Another advantage of using a regulator is that it provides an additional line of defense against any transients on the power line. Any voltage surge which makes it past the varistor, through the transformer, and is not absorbed by the filter capacitor will be largely suppressed by a good regulator. This is especially important in circuits containing flip-flops and one-shots, since a transient may cause them to change states, latching the error.

The introduction several years ago of a complete regulator circuit in a single IC package greatly simplified the design of power supplies with modest current requirements. Generally, in either a TO-220 or TO-3 case, most regulators are three terminal devices. Some of these come with an adjustable output voltage, but ones with standard fixed output voltages are more widely used. In the projects of this book, we have used both the 7805 (a 5 v regulator) and 7812 (a 12 v regulator), which are fixed voltage regulators. The three connections include an input (which goes to the output of the unregulated supply), an output (which supplies the regulated voltage), and a common ground. The schematic symbol for a regulator of this type is shown in figure 2.

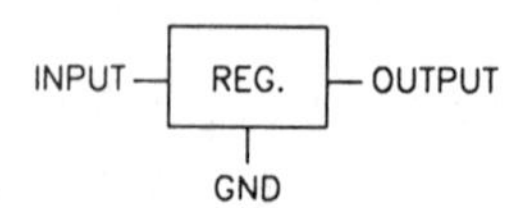

Figure 2 The schematic symbol for a linear regulator IC.

These regulators are basically complete, yet their performance can be enhanced by adding a couple of external components. If the regulator is more than a couple of inches away from the power supply filter capacitor, it is generally considered good practice to include a bypass capacitor (.1 ufd or so) from the input terminal to ground, placed physically close to the IC. Considering how low in cost these capacitors are, and the fact that they can contribute to the circuit's stability, we generally include them regardless of the distance between the regulator and filter. The other addition that can improve circuit performance is to include some additional capacitance on the output of the regulator. This helps the regulator to respond to transients. If the circuit load suddenly increases (perhaps from a relay being turned on), it will take the regulator a certain finite amount of time to respond to the change. An additional capacitance placed on the regulator output can help maintain the proper output voltage during this time.

Whether circuit regulation is maintained by the use of a regulator IC or by using a zener diode and series resistor, no power supply for a circuit employing the use of integrated circuits should be considered complete without including a bypass capacitor on the output. These capacitors, generally in the range of .01-.1 ufd, suppress any transients or RF noise on the power supply lines, and they should preferably be the ceramic disc type. These capacitors perform an important function, appearing as a short circuit to the RF frequencies. In critical circuits, or circuits containing flip-flops or counters, it is also a good idea to place some of these capacitors near the individual ICs to provide local protection. If you are using TTL, the use of these bypass capacitors is even more critical, since the totem-pole output of TTL draws very heavy current from the power supply for a brief period whenever it switches. If you do not include these bypass capacitors, the switching transients can cause all kinds of erratic problems. CMOS is not quite as bad in this respect, but do not try to get by without any bypass capacitors. The circuit may appear to work fine and may never cause you a problem, but, then again, it may work correctly most of the time and only occasionally act erratically. Power supply problems like this can be among the most difficult to track down, so don't take short-cuts. Considering the low cost of these disc capacitors, it makes far more sense to include a sufficient number of them right from the beginning.

DIGITAL INTEGRATED CIRCUITS

The NOR Gate

The 2-input NOR gate is one of the most basic and commonly used digital logic blocks, and is used in several of the projects in this book. Figure 3(a) shows the schematic symbol for a NOR gate with two inputs, while figure 3(b) shows the truth table for this device. Basically, the NOR gate is just an OR gate with an inverter on the output. Looking at figure 3(b), we can see that if either Input A or Input B is high (or both), the output will be low. The output will be high only when both inputs are low.

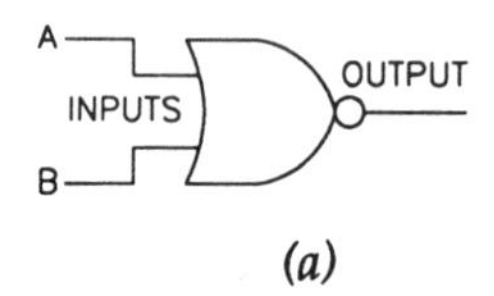

(a)

A	B	OUTPUT
0	0	1
1	0	0
0	1	0
1	1	0

(b)

Figure 3 The NOR Gate (a) The schematic symbol; and (b) the truth table.

The NOR gate is available in both the TTL and CMOS logic families. For the projects in this book, we have used CMOS gates from the CD4000 line exclusively. CMOS tends to have better noise immunity than TTL, and at the low frequencies that these projects operate, much lower current consumption. And while newer and faster CMOS chips are available, the CD4000 family is readily available and low in cost.

For most CMOS gates (and the CD4001 NOR gate, which we use in this book), the switch-point at which the input voltage will cause the output to change states is about one-half of the supply voltage. For example, if you are powering your circuit with a 12 v supply, an input of over 6 v will generally be interpreted as a high input, whereas anything below 6 v will be seen as a low input. Like most CMOS digital gates, however, the inputs of the 4001 should generally be driven as close as possible to either the full supply voltage or ground. This will help to minimize the power dissipation of the IC by always keeping one of the two series FETs in the output stage turned completely off, and also aid in maintaining full noise immunity. Occasionally, an application will come along where these gates are used in a linear mode, with the input operating in between the two voltage extremes. Provided the total power dissipation of the chip is watched, this practice is acceptable, but the current consumption of the IC will increase. For this reason, it is extremely important when using CMOS to tie any unused inputs to either the positive supply or ground.

In addition to providing a logic function, these gates can also be used to form "multivibrators." By cross-coupling two gates, an astable (oscillator), monostable (one-shot), or bistable (flip-flop) multivibrator can be formed. The only difference among the three is the type of feedback used. Figure 4 shows how each of these can be formed. The 2-input NOR gate can also be used as a simple inverter by tying both inputs together, or by connecting the unused input to ground.

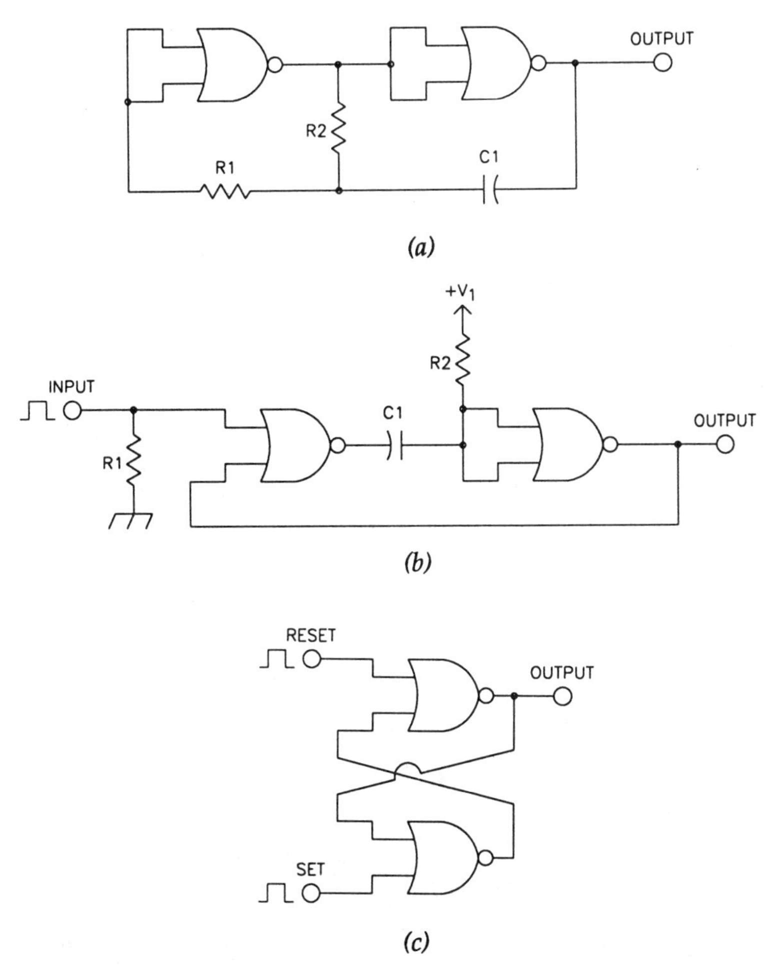

Figure 4 Multivibrator circuits built from simple NOR gates: (a) astable; (b) monostable; and (c) bistable.

The NAND gate

The 2-input NAND gate is quite similar to the 2-input NOR gate, with the exception of its truth table. Figure 5(a) shows the symbol for the 2-input NAND gate, while 5(b) shows the truth table. This gate basically functions as an AND gate with an inverter on the output. Of the projects in this book, the only NAND gate that we use is the 4011, which is also from the CD4000 CMOS family. Its logic level definition is similar to that of the 4001. The 4011 can also be used in multivibrator applications, just as the 4001, with the exception that an 'active low' trigger pulse will generally be needed for the monostable and bistable circuits. It can also be used as an inverter by tying both inputs together or tying the unused input high.

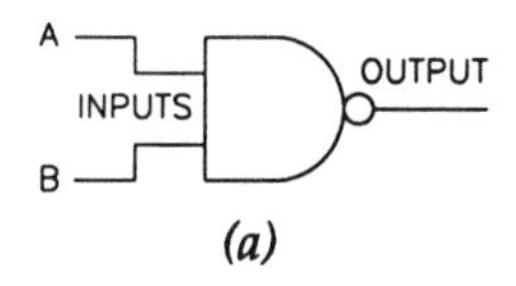

(a)

A	B	OUTPUT
0	0	1
1	0	1
0	1	1
1	1	0

(b)

Figure 5 The NAND Gate (a) The schematic symbol; and (b) the truth table.

The Flip-Flop

The flip-flop is a bistable circuit which can be used as a memory device. Once triggered into a high or low output state, it will remain in that state until another pulse is applied. Used individually, this memory function of the flip-flop can be used to 'remember' whether a relay should be held on or off, based on the last command it received. We use a flip-flop for this function in the Touch-operated Switch and Plant Watering System projects. Several flip-flops can be placed in parallel to form a memory register, which finds extensive use in microcomputer circuits. They can also be cascaded to form counters.

Many different types of flip-flops are available. The three types used in the projects of this book are the Set-Reset, D-type, and the Toggle flip-flops. The simplest of these is the Set-Reset type, which can easily be put together from two gates (see figure 6). This type of flip-flop has two inputs, a "Set" terminal and a "Reset" terminal. Using NOR gates, as shown here, both of the inputs are normally held low, going high only when the flip-flop is to be triggered.

Looking at figure 6, we can see that if pin 1 (the "Set" terminal) is brought high, pin 3 must go low, since the output of a NOR gate must be low if either of its inputs is high. When pin 3 goes low, pin 5 is also brought low. Since pins 5 and 6 are both low, pin 4 will be high; thus, Q goes high when a positive pulse is applied to the "Set" terminal. In addition, pin 4 going high also holds pin 2 high, which, in turn, holds pin 3 low even after the "Set" trigger pulse is gone. Our circuit has therefore latched a high state on the Q terminal.

If the "Reset" terminal is brought high, however, Q is forced low. Pin 2 is also brought low in the process. With pins 1 and 2 both low, pins 3 and 5 go high, latching pin 4 low. The circuit will now be latched with a low state on the Q output. Note that in both states, a complementary, or inverted, output is also available on pin 3.

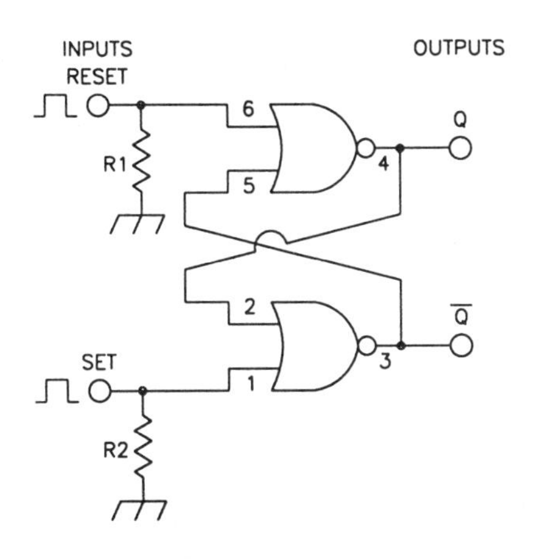

Figure 6 A set-reset flip-flop.

The D-type flip-flop is a form of clocked logic (its schematic symbol is shown in figure 7). To use this device, the logic level to be latched on the Q output is placed on the 'D' ("Data") terminal; the output, however, doesn't respond immediately. The data on D will not be latched on Q until the clock terminal ("C") is pulsed high. Note that unlike the Set-Reset flip-flop, the D-type is edge-triggered. The actual logic level of the clock terminal is unimportant. It is the transition from low to high on the clock that causes the data to be latched on Q; thus, if the clock terminal is brought high and remains high, only the value on D at the moment the clock went high is transferred to Q. The high level on the clock terminal after the transition has no effect on the output; however, a level-triggered "Reset" terminal is usually included. This input can be used to preset the flip-flop to a known state and generally overrides all other inputs. Several D-type flip-flops can be combined in groups to form memory registers, widely used in computer circuitry. They can also be cascaded to form shift registers and sequential logic circuits. In the projects included here, we have used the CD4013 D-type flip-flop to build monostable ("one-shot") circuits, and to construct Toggle flip-flops.

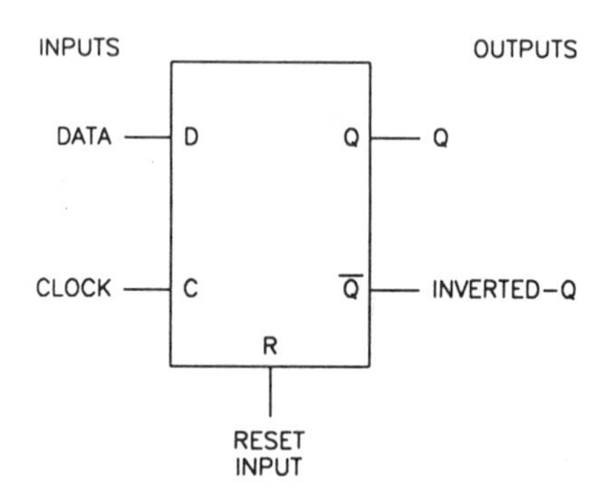

Figure 7 A D-type flip-flop.

Although the Toggle (or "T") flip-flop is available in IC form, it is more common to form it from more popular ICs. The Toggle flip-flop has only one input, to which clock pulses are applied. Each pulse applied will cause the output to change states, so that it alternates between high and low with every clocking. This flip-flop can, therefore, be used as a frequency divider (since it alternates states, it will divide an incoming pulse train by two) or cascaded to form a binary counter. It can also be used, as it is in this book, as an alternate-action switch, turning a relay on and off on alternate inputs.

A T-type flip-flop can be easily made from a D-type by simply connecting the D input to the inverted-Q output. Since this output is the complement of Q, tying it to D will cause Q to alternate states with every clock pulse that is applied. We use this technique to generate the toggle action in the 'Touch-operated Switch' project.

The Counter

Counters are made by interconnecting several flip-flops. Many different types exist, ranging from the simple binary counter to more elaborate Up/Down Synchronous counters. The only counter used in this book is the CD4017. We have described its operation in detail in the 'Light Sequencer' project.

We have attempted to make these projects as educational as possible, and we hope that you will find them to be fun. As necessary as they are, textbooks on electronics theory can get a little dry at times. It is here, where you learn how to utilize this knowledge to do useful things, that you achieve a sense of accomplishment. It is our sincere hope that you will not only learn from these projects, but that you will find them to be enjoyable and satisfying.

ELECTRONIC THERMOSTAT

This project is an electronic thermostat which can be used for controlling the temperature of a device or detecting temperature change. It can be used with various heating and cooling devices for applications such as controlling house or attic fans, incubators, electric heaters, or for detecting temperature changes, such as an alarm to indicate that the temperature in a freezer is rising. This circuit, when constructed as described here, is capable of switching 13 amp at 120 v, over a temperature range of about 60-110°F. Other temperature ranges can be obtained by simply changing a resistor value. Construction time will vary depending upon the application and the choice of enclosure, but it can easily be assembled over a weekend.

Exterior view of a completed electronic thermostat project.

CIRCUIT DESCRIPTION

Our application for this device was to automatically turn a house fan on and off, while still providing manual operation when desired (see figure 1 schematic). Power to the project was furnished by a standard line cord. The neutral (white) wire from the cord connects to the neutral wire exiting to the controlled device. The hot (black) wire connects to the common (center) terminal of the power switch. We used a DPDT switch with a "center off" position (Radio Shack model #275-1533) and shorted the terminals for greater reliability (see figure 2).

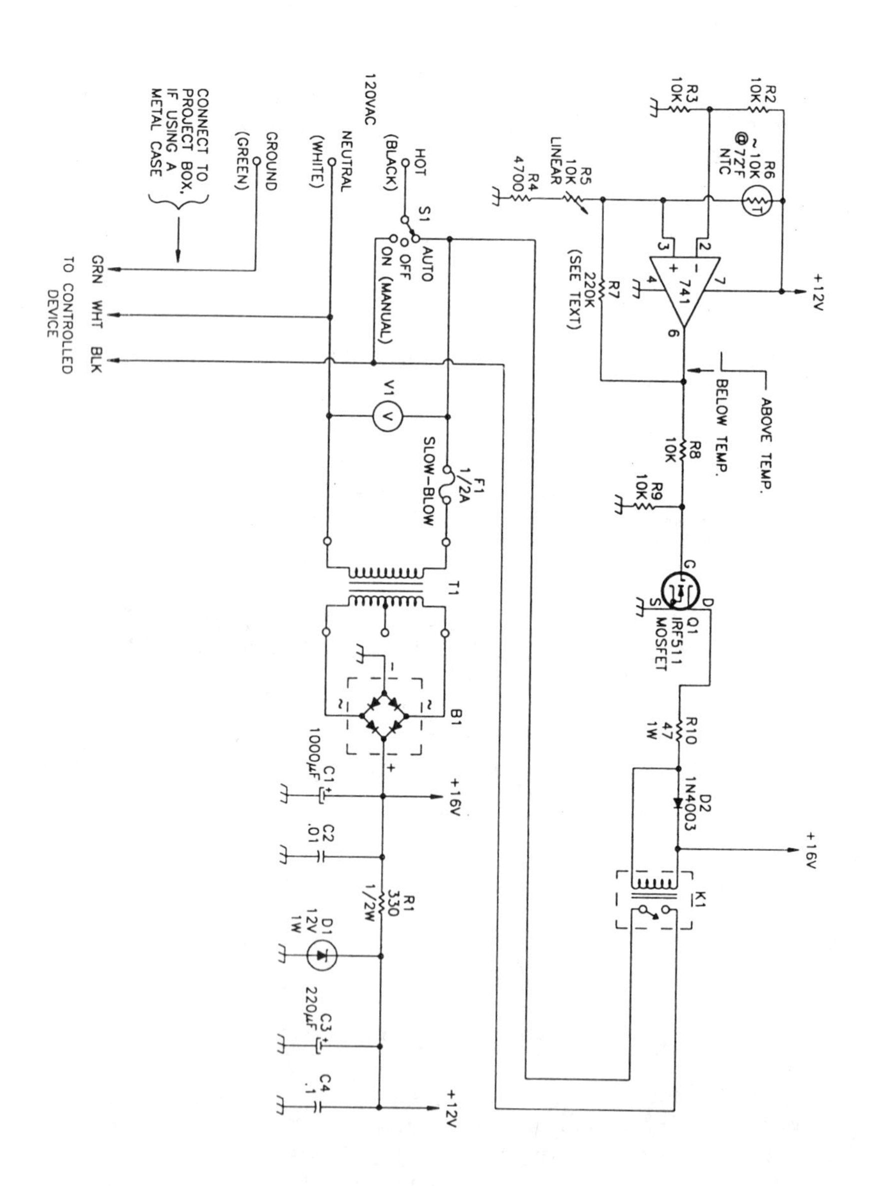

Figure 1 Schematic of the electronic thermostat.

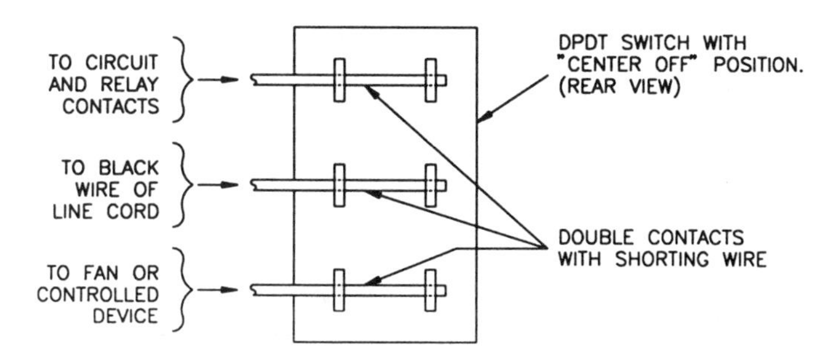

Figure 2 Doubling the contacts on the switch for greater reliability.

In the "center" position, the switch is open and the power is disconnected. In the "manual" position, power is applied directly to the device. In the "auto" position, power is applied to the normally open contacts of power relay K1 and to transformer T1 to power the circuit. If your application does not require manual operation of the device, a SPST switch can be substituted for simple on/off control, or it can be eliminated entirely. The varistor, V1, is not essential to circuit operation, but it helps to suppress any transients coming over the power line or generated by the switched device. Note that fuse F1 does not protect the controlled device, only the thermostat circuit.

The output of the transformer provides 12 vac to the rectifier bridge, which converts the signal to a pulsating DC voltage, filtered by capacitor C1. Capacitor C2 provides RF bypassing. This voltage is fed through resistor R1 to zener diode D1, which regulates the DC voltage at +12 v to power the 741 op-amp. Capacitors C3 and C4 provide additional filtering and bypassing.

The +12 v are used to power the control circuit, consisting of the 741 op-amp, thermistor R6, potentiometer R5, and resistors R2, R3, R4, and R7. R2 and R3 form a voltage divider, which holds the inverting (-) terminal of the 741 at about 6 v. Thermistor R6 forms another voltage divider with R4 and potentiometer R5, which connects to the non-inverting (+) terminal of the 741. The voltage on the non-inverting input will vary with the setting of the potentiometer and the value of the thermistor's resistance at any given temperature. Since the thermistor will decrease in resistance with a temperature increase, this voltage will tend to rise as the temperature goes up. When the voltage exceeds the level set on the inverting terminal, the output of the 741 will swing positive. Note that the op-amp is running "wide open." Because there is no negative feedback to limit the gain, the output of the 741 will swing fully positive when the temperature rises above the set point (to about 10-11 v) and will swing fully negative when the temperature is below the set point, to within a volt or two of ground.

Resistor R7 provides positive feedback from the 741 output to the non-inverting input. Its purpose is to furnish a certain amount of hysteresis. By this, we mean that the temperature at which the circuit switches will be different when the temperature is rising than when it is falling; therefore, there will be a differential between the cut-off and cut-on temperatures.

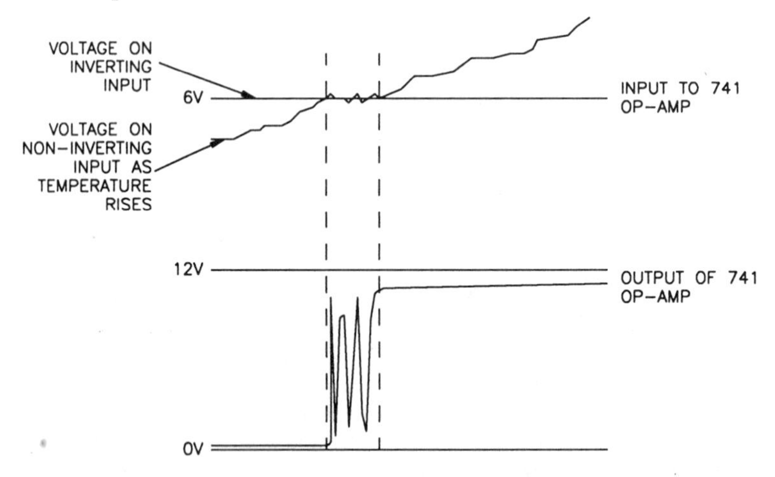

Figure 3 The output of the 741 may hesitate if no hysteresis is provided, due to the slowly rising voltage on the non-inverting input.

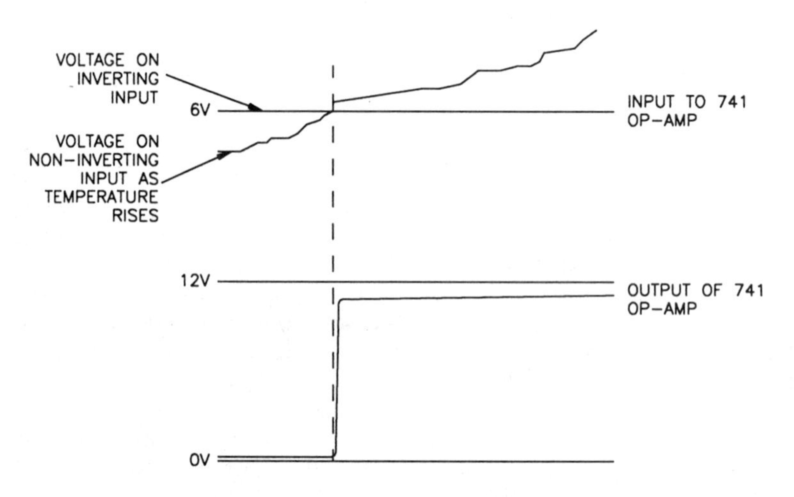

Figure 4 The addition of hysteresis produces a "snap-action" that totally eliminates any output hesitation.

This differential is normally desirable because of two reasons. The first reason is that the resistance of the thermistor generally changes very slowly, since most temperature changes are gradual. At the point where the + and - inputs of the op-amp are nearly equal, the output can turn on and off rapidly, causing the relay to "chatter," possibly damaging the controlled device (see figure 3). The positive feedback provided by R7 gives a "snap-action" effect to eliminate this problem (see figure 4).

To understand how this works, consider that the output of the op-amp is always near +12 v or ground. With the output low, R7 can be thought of as in parallel with R4 and R5 (see figure 5). When the output switches high, R7 is in parallel with R6. The instantaneous decrease of resistance in the upper portion of the divider, accompanied by the increase of resistance across R4 and R5 with the removal of R7, causes a sudden voltage level shift on the non-inverting input. In the example of figure 5, with a 100K resistor for R7, the voltage on the non-inverting input is "pulled up" by over 1/2 v as the output switches. The resistance of the thermistor will now have to increase by a substantial amount before the circuit can switch back to the low state.

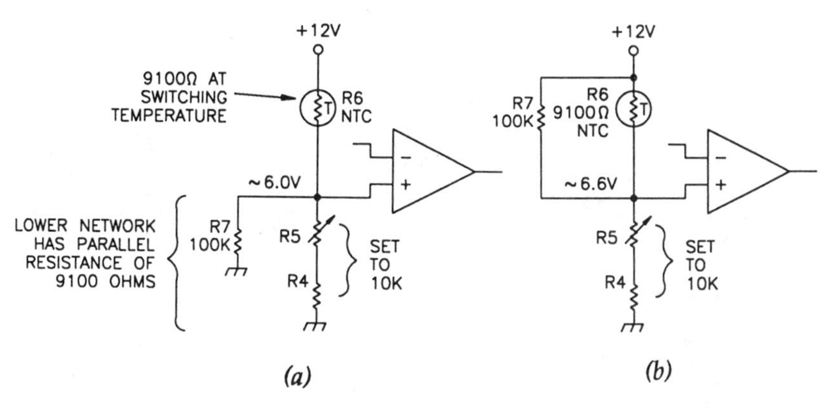

Figure 5 (a) The equivalent circuit just before the circuit switches from low to high; and (b) the circuit after the output has gone high. Although it is helpful to think of the "switching" end of R7 as being at either ground or +12 v, the output of the 741 never gets to within a volt of either one. The actual hysteresis will, therefore, be lower than the calculated value.

Besides preventing output instability, the second reason that we want a differential between the cut-on and cut-off points is to prevent unnecessary wear on the power relay and the controlled device. Only a very small amount of hysteresis is necessary to prevent the output chatter problem just discussed, but it is often desirable to introduce several degrees difference between the turn-on and turn-off points in many applications. For instance, suppose an attic fan were set to come on at 100°F and had a differential of only 1/10°F. As the temperature rose to

100°F in the attic, the fan would turn on. As the fan pulled in cooler air from outside, the temperature might drop 1/10°F in only a matter of seconds, turning the fan back off. As the fan stopped, the hot stagnant air in the attic would soon raise the temperature enough to cycle the fan on again; thus, the fan might cycle on and off hundreds of times a day. Lowering the cut-off point several degrees solves this problem. This can be done by varying the resistance of R7; the lower the value, the greater the differential.

When the output goes high on the 741, approximately 5-6 v are applied to the gate of Q1. This turns on the FET, which energizes the relay, turning on the device. Since K1 is a 12 v relay, and the unregulated voltage is about 16 v, R10 was added to insure that the coil's wattage rating is not exceeded.

CONSTRUCTION DETAILS

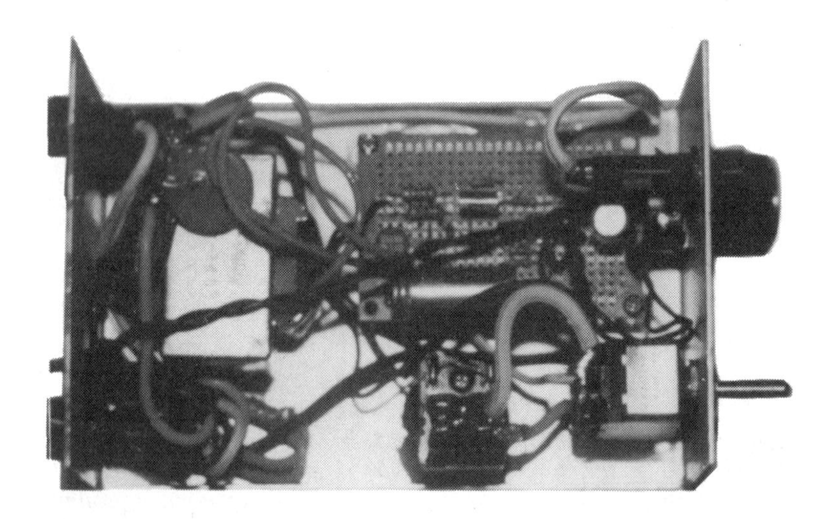

Interior view of a completed electronic thermostat project.

The control circuitry for the prototype was constructed on a pre-etched circuit board available from Radio Shack, but any standard construction technique can be used. Since this unit may be used to control a device of high voltage and high amperage, adequate safety precautions should be taken. If a metal enclosure is used, make certain to use a three-wire electrical cord with the ground wire (green) attached to the case. For our unit, we found it convenient and economical to simply take a three-wire extension cord with molded ends and cut it. This provides both the power cord and the device power interface. The AC wiring inside the box should be of sufficient gauge to handle the current requirements of the controlled device. By using 16 gauge wire along with the switch and relay in the parts list, 13 amp can be controlled with this device. Remember to take into account start-up current on large motors.

The relay used is a DPDT model. It is a good idea to double the contacts together, as mentioned earlier regarding the switch, to increase the current carrying capacity. Since the relay had no mounting bracket, we mounted it with silicone rubber.

The thermistor must be outside of the case for reliable operation, because the circuit itself will generate a certain amount of heat. We soldered leads to the thermistor and brought them into the case through a grommet. For many applications, the thermistor can simply be suspended by its leads a couple of inches from the box. If the thermistor must be mounted a distance away from the control circuitry, it might be a good idea to add a .1 ufd disc capacitor across the thermistor leads where they attach to the circuit board in order to suppress any noise induced on the lines.

The circuit as shown in the schematic is designed to be used to detect a rising temperature, as in cooling applications. A rise in temperature will cause the device to turn on. If the circuit is to be used for heating applications, reverse the position of the thermistor in the circuit with R4 and R5 (see figure 6). This will turn the controlled device on as the temperature drops.

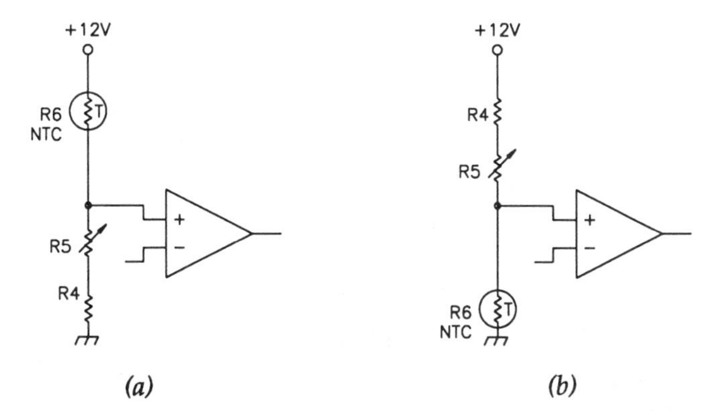

Figure 6 (a) The original circuit, wired to turn on as the temperature rises; and (b) the circuit rearranged to turn off as the temperature rises, for heating applications.

The temperature range of the circuit shown is approximately 60-110°F. This range can be shifted by changing R4. Lowering R4 raises the temperature range and increasing R4 causes the temperature range to drop. Note that with potentiometer R5 at mid-range, R4 and R5 are about equal to the resistance of the thermistor at room temperature. For another range, use the chart provided with the thermistor to find its resistance value at the temperature desired, and then select R4 and R5 to match this with R5 about mid-range.

The cut-on/cut-off differential is determined by the value of R7. If very tight regulation is needed, this value should be relatively high. Ten megohms will generally provide enough hysteresis to prevent output chatter, but lower values are better for most applications. A value of 1 meg should provide regulation of about 1/2°F, a good choice for an incubator. For general purpose applications, a 220K resistor is probably ideal, usually giving about a 2°F differential. For attic fans, even more hysteresis is desirable. A 100K resistor will provide a 4-5°F differential, about right for such applications. For use as a temperature detector, such as a freezer alarm, the hysteresis is of no real value to the application, and R7 can be in the range of 1-10 meg. For applications requiring very precise control of the differential, or where the needed differential may change, a potentiometer can be substituted for R7.

If you are using this as a freezer alarm, or in any application where the temperature is very high or low, it is best to place the thermistor in the environment to be sampled and attach it through longer leads to the thermostat, rather than placing the entire circuit in a hostile environment. For temperature alarm applications, relay K1 can simply be replaced by a 12 v buzzer, and R4 or R5 can be changed if the desired alarm temperature is not in the 60-110°F range. For freezer alarms, a 15k or 22k resistor for R4, a 50K linear potentiometer for R5, and a 1 meg resistor for R7 would be a good combination.

PARTS LIST

Semiconductors:

B1- 100 PIV-1.2A., diode bridge (Radio Shack #276-1152)
D1- 1N4742-12 v-1 w zener diode (Radio Shack #276-563)
D2- 1N4003-rectifier (Radio Shack #276-1102)
Q1- 1RF511 FET (Radio Shack #276-2072)
U1- 741 op-amp (Radio Shack #276-007)

Capacitors:

C1- 1000 ufd-35 v electrolytic
C2- .01 ufd-disc ceramic
C3- 220 ufd-16 v electrolytic
C4- .1 ufd-disc ceramic

Resistors: (All resistors 1/4 w unless stated otherwise)

R1- 330 ohm-1/2 w
R2, R3, R8, R9- 10K-1/4 w
R4- 4.7K-1/4 w
R5- 10K linear potentiometer
R6- thermistor (Radio Shack #271-110)
R7- 220K-1/4 w (see text)
R10- 47 ohm-1 w (or 2-100 ohm-1/2 w, in parallel)

Miscellaneous:

K1- power relay (Radio Shack #275-218)
T1- 12 v transformer, 450 mA (Radio Shack #273-1365)
V1- varistor (Radio Shack #276-570)
DPDT switch (Radio Shack #275-1533) **Optional
Fuse holder .5A slow-blow fuse
Power cord
Enclosure
Pilot light **Optional
Circuit board (Radio Shack #276-150)

BRINGING UP THE UNIT

After constructing the electronic thermostat and verifying that it is wired properly, plug the unit in and plug a lamp or other test load into its receptacle. Turn the temperature control fully clockwise (counter-clockwise if the device has been wired for heating rather than cooling applications). Move S1 to the "auto" position and make sure the load is off. Turn the temperature control counter-clockwise (for cooling), and the relay should switch on when the control is about mid-range. Turning the control back the other way should turn it back off; however, there will be a noticeable difference in the turn-on and turn-off points. This is due to the hysteresis introduced by R7. If you have wired the device to turn on when the temperature rises, as shown in the schematic, and the relay energizes when the knob is turned clockwise, you have wired the potentiometer backwards.

If the unit does not seem to function properly, check the following:

- With the switch in the "auto" position, carefully check the voltage across capacitor C1. The voltage here should be a minimum of 14 v; a reading of 16-19 v is typical. If the voltage is lower than this, or zero, check the wiring of T1 and the diode bridge. Make certain that 120 v is on the primary of T1.

- If the voltage in step #1 is correct, measure the voltage across pins 7(+) and 4(-) of the 741 op-amp. There should be about 12 v here. If this is not correct, check the wiring of R1 and the zener diode. Make certain that the zener has been installed with the positive terminal connected to ground.
- If all power supply voltages are correct, momentarily short the gate of the FET to +12 v. The relay should energize. If it does not, check to make certain that power is getting to the relay, and that the diode is wired as shown. Make certain that the source of the FET has a good ground connection, and that it has been wired correctly. If shorting the source and drain causes the relay to energize, but putting +12 v on the gate does not, replace the FET.
- If the relay energized correctly in step #3, the problem is in the control circuit. Measure the voltage at pin 2 (the inverting input) of the 741. About 6 v should be here. If not, there is a problem with the wiring of R2 and R3. If this is correct, measure the voltage at pin 3 of the 741 (the non-inverting input). As you rotate the temperature control knob, the voltage should change, but at some point, the voltage should cross +6 v. If not, make certain R4 and R5 are the correct value and wired properly. If the voltage will not reach +6 v, turn off the power and measure the resistance of the thermistor with one lead removed from the circuit. At room temperature, the resistance should be about 10K ohms. If it is not, replace the thermistor.
- If the voltage on the non-inverting input does go above the +6 v on pin 2, but the relay does not energize, check the output of the 741 (pin 7). If it does not go high (10-11 v) when pin 3 is more positive than pin 2, replace the 741. If it does go high, check the connections on R8 and R9.

LIGHT SEQUENCER

The Light Sequencer is a simple, easy to build, yet highly educational project that can be assembled in a very short time. One or more lights (LEDs or incandescent) can be attached to each of its ten available outputs, which are turned on one at a time in sequence. By connecting several lamps to each output and connecting them in a ring, a light "chaser" can be created. Other uses for this circuit include advertising displays, sequential turn signals, and toys. Although driving lights serve as a visual aid in understanding it, this type of circuit also finds wide use in other applications, as in stepper motor and sequential logic circuits. This project demonstrates the operation of a sequential state machine in an easy to understand form.

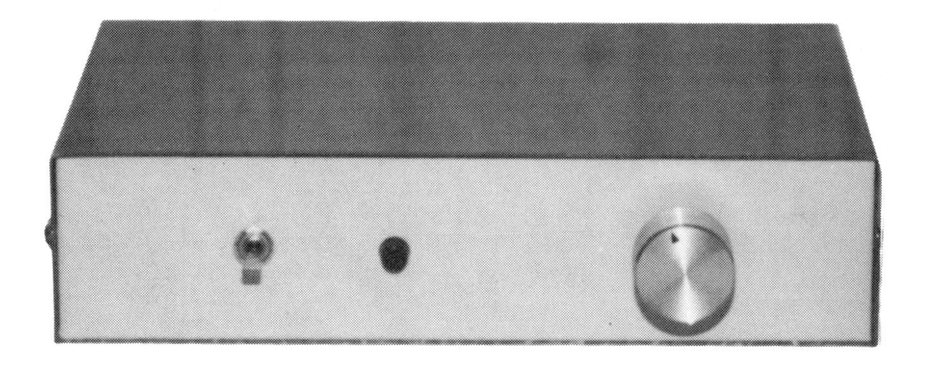

Exterior view of a completed light sequencer project.

CIRCUIT DESCRIPTION

In order to build a light sequencer, we will need a circuit consisting of two stages: (1)a "stepper" stage with several "1-of-n" outputs, with "n" being the number of steps in the sequence. If you are building a chaser display where every fifth LED will be lit, then "n" will be 5. The stepper must have only one output active at a time and must sequence to the next output upon receiving a clock pulse, and (2) a "clock" stage, to generate clock pulses for the stepper circuit at regular intervals. The frequency of the clock will determine how fast the stepper sequences, so the frequency output of this stage should be adjustable. In addition, we will also need a power supply. Figure 1 shows the entire light sequencer circuit used in this project containing all of these components.

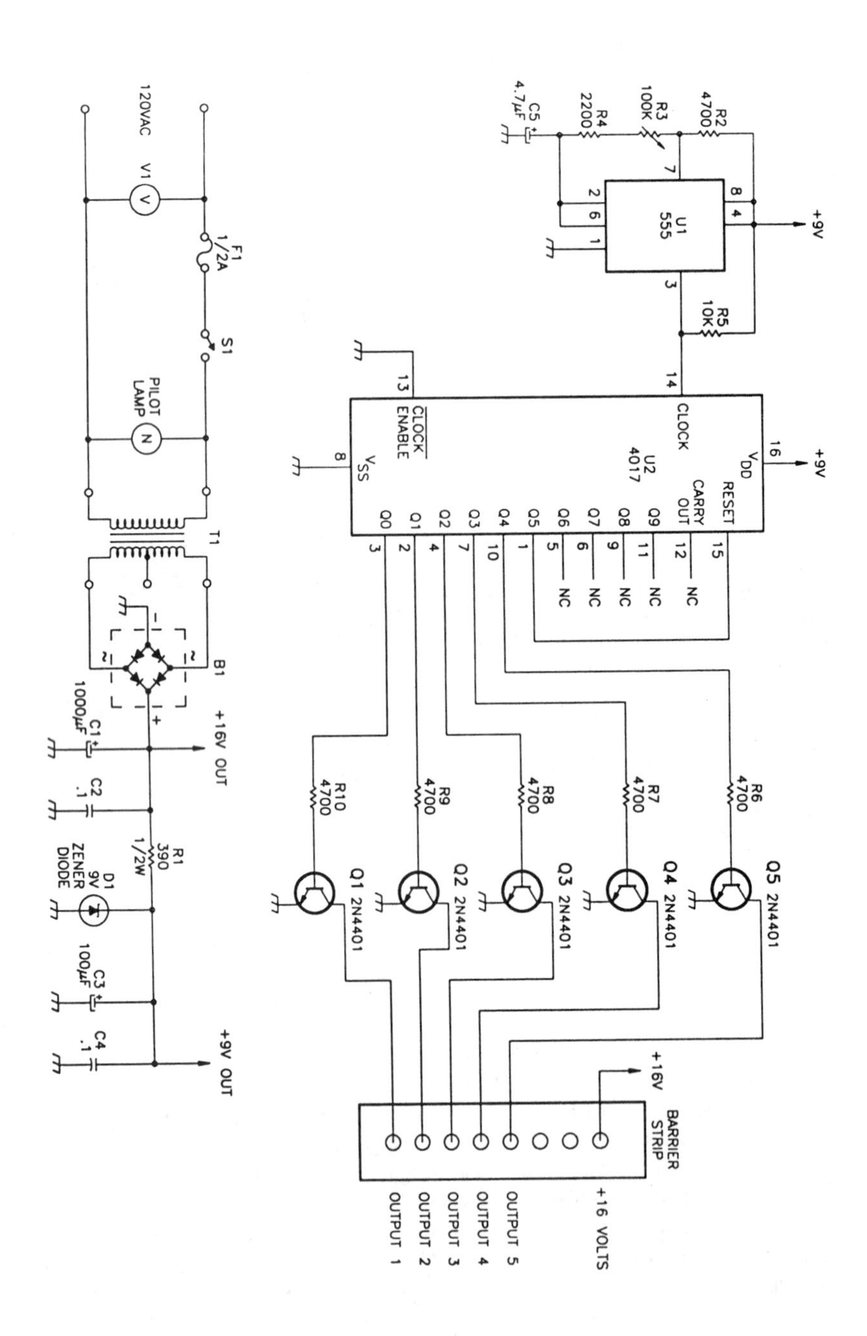

Figure 1 Schematic of the 'Light Sequencer' project.

The stepper circuit for this project is shown in figure 2. It consists of a 4017B IC driving several transistors, one for each output being used. This chip is a decade counter with a "1-of-10" output code. Unlike many counters, which have a binary output, this counter has a separate output for each of its ten possible states. The first output, QO, is high when the internal count is zero. The other nine outputs are low. When a clock pulse is received on the "clock" terminal, the counter will increment. Output QO then goes low, and output Q1 goes high. The next clock pulse causes Q1 to go low, and Q2 to go high, and so on. Normally, when Q9 is high and a clock pulse is received, the counter resets to zero and starts over, but for our application, we have shortened the count length by tying Q5 to the reset terminal. When Q5 goes high, it will cause a reset, and the cycle will start over after only five increments.

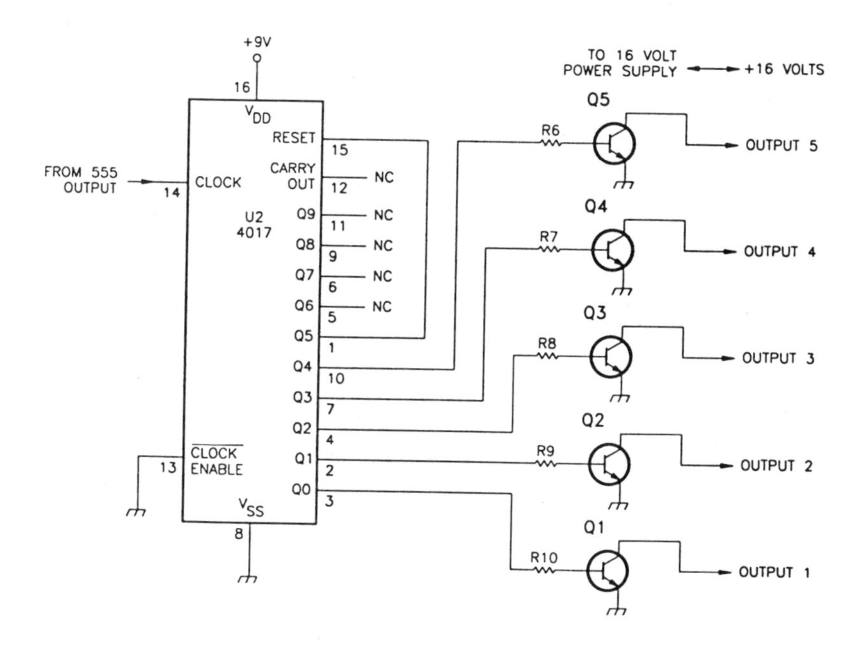

Figure 2 The stepper circuit of the light sequencer.

The 4017 decade counter is perfect for our application here, and is widely used in this and similar "stepper" tasks. Internally, it is configured as a Johnson, or "walking ring" counter. One of the primary advantages of this configuration is the ease with which "1-of-n" outputs can be decoded. To better understand what is going on inside this IC, take a look at figure 3.

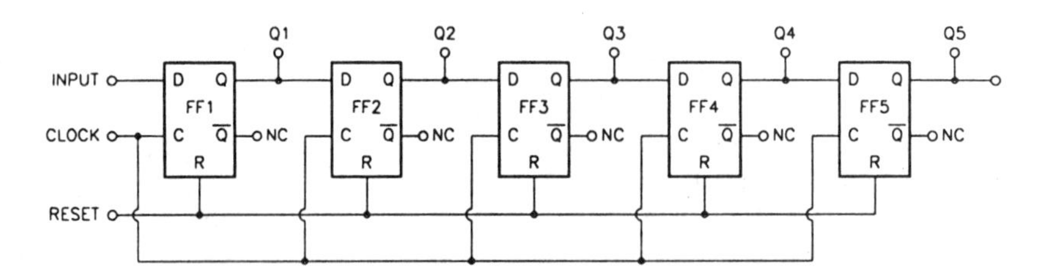

(a) A "bucket brigade" circuit. Applying a clock pulse will result in the data on each output being transferred to the flip-flop on its right. The value on the input terminal will appear on Q1 after clocking, but the data on Q5 before will be lost.

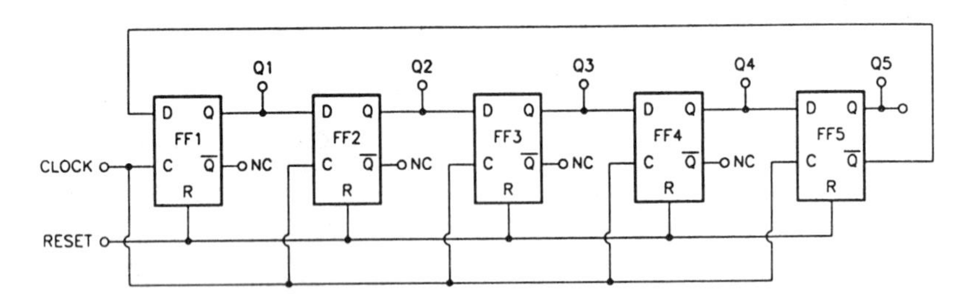

(b) A "walking ring" counter with ten states

Figure 3 Conversion of a "bucket brigade" circuit into a "walking ring" counter.

In figure 3(a), we show a typical "bucket brigade" circuit, which is at the heart of most stepper and shift register designs. We have used five D type flip-flops in this particular circuit, similar to the 4013. You will recall that this type of flip-flop, upon receiving a clock pulse, will take the logic value on its D terminal and transfer it to its Q output, latching it there until a new clock pulse or reset signal is received. In this circuit, the D input of each flip-flop is connected to the Q output of the flip-flop on its left, except for the first one, whose D terminal serves as the input. Since the value on the D input will be transferred to the Q output on a low to high transition of the "clock" terminal, a clock pulse applied to this circuit will cause the value on our circuit input to appear on terminal Q1. Notice, however, that the "clock" terminals of all five flip-flops are tied together; therefore, at the same time that the input value is being transferred to Q1, the value previously on Q1 is being transferred to terminal Q2, the value previously on Q2 is being transferred to Q3, and so on. The value on terminal Q5 before the clock pulse will be lost after clocking; therefore, a value placed on the circuit input will be latched on Q1 with the first clock pulse and sequentially moved down the line with each additional clock pulse.

If we were to preset the five flip-flops to a certain value, such as 10000, and connect the Q5 output to the circuit input, we could close the sequence on itself and make a sequential loop, which would repeat cycle after cycle. The "walking ring" counter is a slight variation of this, in that we take the inverted-Q output of the last flip-flop and connect it to the input. Although slightly more difficult to decode, this has the advantage of giving us ten output states, rather than five, for the same number of flip-flops used. It also has the advantage of including state 00000 in its sequence. This simplifies the preset requirements, since a simple reset pulse automatically puts the counter in a valid state.

Figure 3(b) shows a typical "walking ring" configuration with ten states. To understand how the count progresses, let's assume our counter currently has a count of zero. Since all five Q outputs are low, the inputs to the D terminals are also low, except for FF1, which has its D input connected to the inverted-Q output of FF5. On the arrival of the first clock pulse, the count will therefore change from 00000 to 10000. Now, both FF1 and FF2 have a high on their D inputs; therefore, the next clock pulse will advance the count to 11000. After five clock pulses, the count will be 11111, but as output Q5 goes high, the inverted-Q output of FF5 goes low. The sixth clock pulse will therefore advance the counter to 01111. The next pulse will result in an output of 00111, and so on, until the tenth pulse arrives, returning the count to 00000 and starting the cycle over. Figure 4 shows a summary of the count sequence.

Counter State	Walking-ring Output Sequence Q1	Q2	Q3	Q4	Q5
0	0	0	0	0	0
1	1	0	0	0	0
2	1	1	0	0	0
3	1	1	1	0	0
4	1	1	1	1	0
5	1	1	1	1	1
6	0	1	1	1	1
7	0	0	1	1	1
8	0	0	0	1	1
9	0	0	0	0	1

Figure 4 A summary of the "walking ring" output sequence.

As we stated previously, the "walking ring" configuration allows for very easy "1-of-n" decoding. Unlike a binary counter, any of the ten possible outputs can be decoded from the ring with a simple 2-input AND gate. Connecting one input of the gate to Q1 and one to Q5 decodes the number `5,' since this is the only point in the sequence when both Q1 and Q5 are high. Connecting the two inputs of the AND gate to the inverted outputs of Q1 and Q5 decodes the number `0,' since this is the only point in the sequence when both Q1 and Q5 are low. Figure 5 shows how all ten states can be decoded from the "walking ring."

Although not terribly important to our application, another advantage of this type of counter is the fact that it is synchronous. All outputs change simultaneously, and no output glitches are produced, as with a long binary counter. Because all stages are checked at the same time and only one output in the ring changes per clock pulse, this circuit is inherently syncronous and glitch-free.

One disadvantage, however, is that only ten states are used with five flip-flops, while a binary counter of the same length has thirty-two usable states. The problem is not just the lack of efficiency; care must be taken to insure that the counter does not accidently get into one of the twenty-two unused states. If it does, you will have problems. An example for our "walking ring" in figure 3(b) might be the value 10110, which should never appear in normal operation. These unwanted values, known as "disallowed states," must be prevented. In the case of the 4017, however, this has already been taken care of. It includes gating circuitry to force it out of any wrong states and back into its normal count sequence.

Since the 4017 has ten outputs, up to ten separate "steps" of lights can be connected. If more than ten lights are wanted, such as for a light chaser, you can connect several lights to each output, laying them out so that every tenth light is connected to output 0, every tenth one is connected to output 1, etc. Normally, however, you will not need all ten outputs. Even with a chaser display, you will only use five or six. We can shorten the count length of the 4017 by tying the first unused output to the reset pin. If we are using five outputs (Q0-Q4), we can tie the sixth output, Q5, to reset. The first five clock pulses will cycle through outputs Q0 to Q4. The next pulse will set output Q5, but when this output goes high, it will immediately force a reset, causing Q0 to go high and starting the cycle over.

As each output goes high, it turns on its respective output transistor, which provides a path to ground for the lamps on that output. Although we show a 2N4401 transistor, a TIP120 Darlington transistor or IRF511 MOSFET can be substituted if you will be driving heavy loads (over 150 mA). Either of these are available at Radio Shack, and both will handle at least 3 amp. For very heavy loads, heat sinking may be necessary.

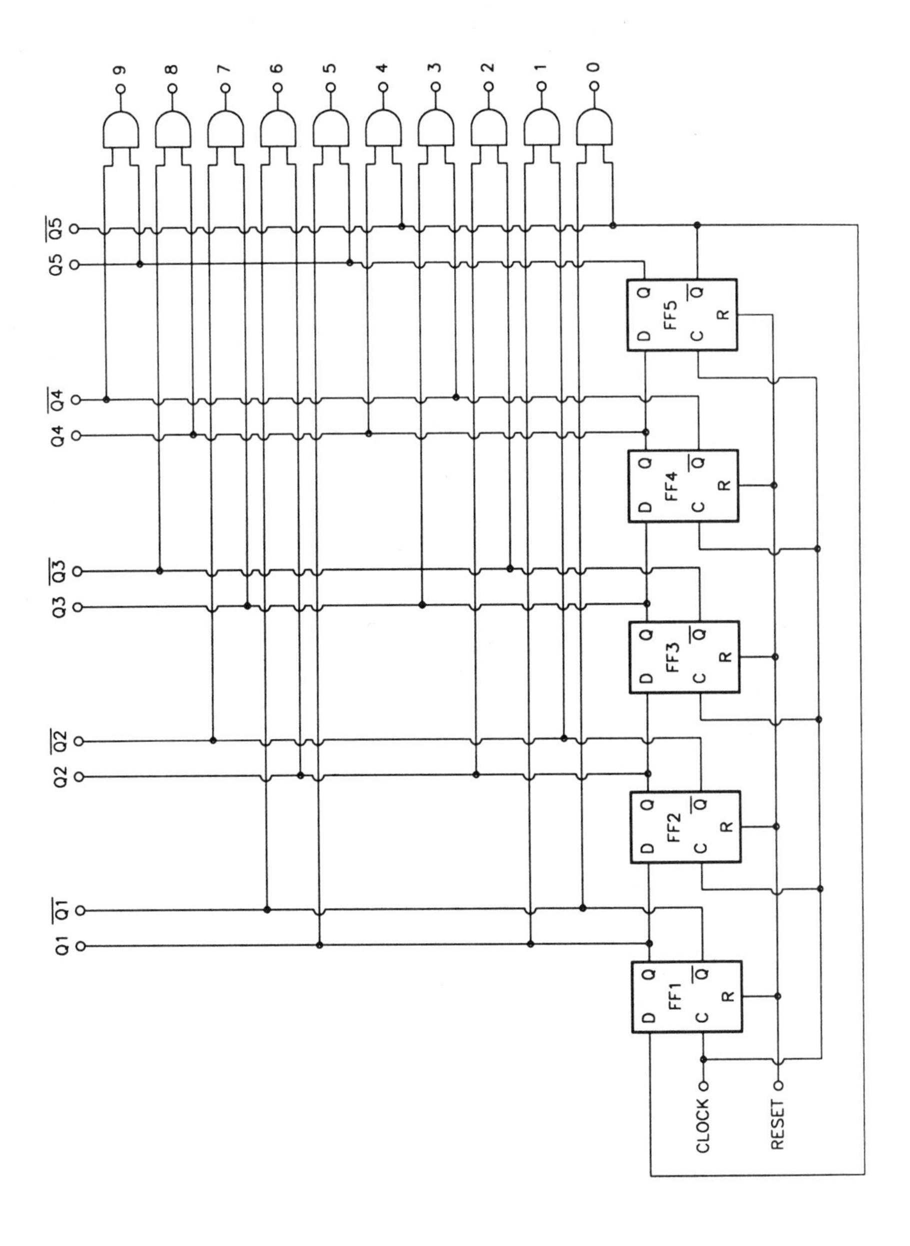

Figure 5 Decoding "1-of-n" outputs from a "walking ring."

Figure 6(a) shows the clock pulse generator. It is a simple astable multivibrator built around a 555 timer IC. If you have built many electronics projects before, you have almost certainly used this versatile chip.

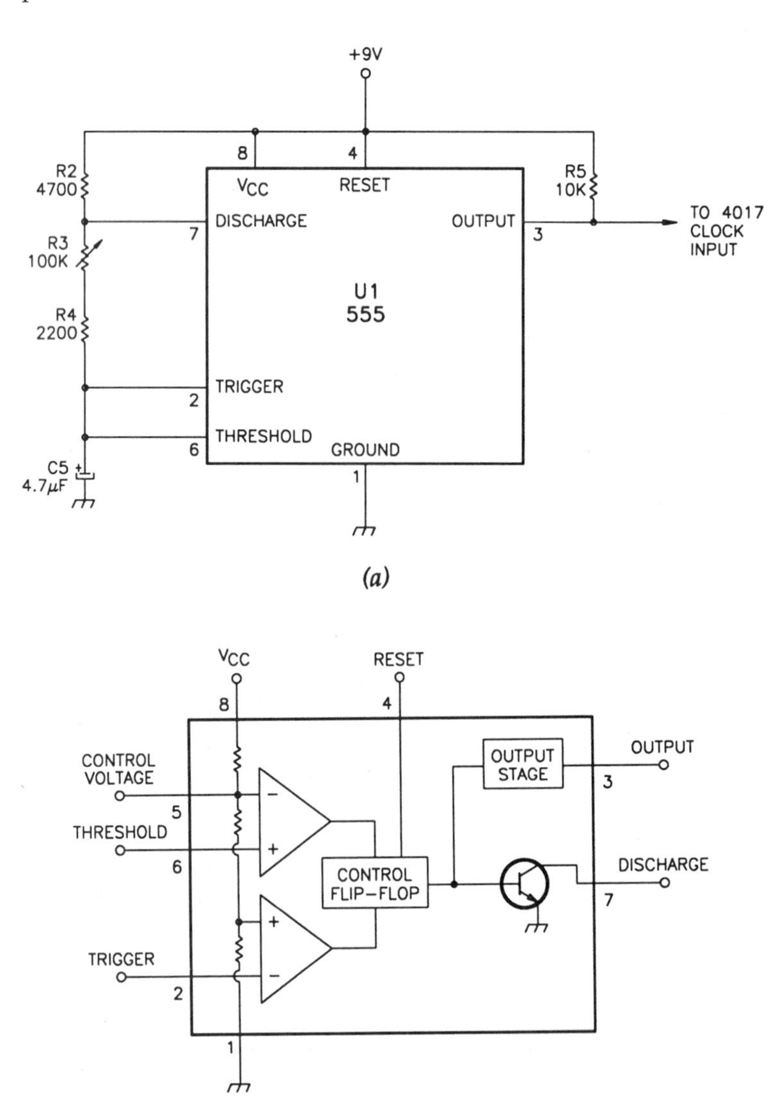

Figure 6 (a) The clock pulse generator circuit; (b) the internal structure of the 555.

If you are not familiar with the 555, figure 6(b) shows its internal structure. Two comparators and a flip-flop are arranged so that the flip-flop is set (causing pin 3 to go high) whenever the voltage on the trigger terminal (pin 2) goes below one-third of the supply voltage, provided the reset pin is held high (its inactive state). Once the flip-flop is set, pin 3 will stay high until the voltage on pin 6 (the threshold terminal) goes above two-thirds of the supply voltage or until a reset pulse is applied. When the flip-flop is reset, the output goes low, and the transistor connected to pin 7 turns on to discharge the timing capacitor. If the one-third and two-third levels need to be altered, pin 5 can be used to set an alternate trigger and threshold voltage.

To use this IC in an astable mode, connect it as shown in figure 6(a). Pin 4, the reset, is tied to pin 8 to hold it high. Unlike CMOS devices, the 555 uses the TTL convention of an "active low" signal; that is, the pin must be brought low to cause a reset. Pin 3, the output, is connected to the clock terminal of the 4017 stepper. On power-up, pins 2 and 6 of the 555 are held low by capacitor C5. Since pin 2 is below 3 v (one-third of the supply voltage), the internal flip-flop is set. Pin 3 will go high, and the discharge transistor connected to pin 7 is turned off, but as C5 charges through R2, R3 and R4, pin 6 eventually will have a potential of 6 v (two-thirds of the supply voltage) on it. This will cause the internal flip-flop to reset. Pin 3 will go low, and the discharge transistor will turn on. At this point in time, pin 7 will be nearly at ground potential. C5 will then start to discharge through R3 and R4. When the voltage on C5 falls to 3 v, the flip-flop will again be set, and the cycle will repeat; we, therefore, end up with a squarewave oscillator whose frequency is determined by the time constant of C5 in combination with R2-R4. The actual frequency of oscillation can be found by the formula 1.44/(Ra+2Rb)C, where Ra is R2, and Rb is R3+R4.

As we lower the value of R3, C5 will charge and discharge faster, increasing the frequency of the signal source. This will cause our stepper action to speed up. Increasing the value of R3 slows the charging time of C5, therefore slowing down the stepper circuit.

The purpose of R5 is to act as a "pull-up" resistor. CMOS chips, such as the 4017, prefer for their inputs to be driven to nearly the full supply voltage, or from fully to ground. The 555, being a bipolar device, pulls very close to ground on a low output, but cannot raise its output to the full supply voltage without some outside help; therefore, we insert R5 to pull the voltage the remainder of the way up on a high output. A CMOS version of the 555 is available (Radio Shack #276-1718) whose output will go nearly from one supply rail to the other. If you want to pay a few cents more for this part, R5 can be eliminated.

The power supply, as shown in figure 1, is designed to run off the AC power line. Transformer T1 drops the AC voltage to about 12.6 vac. Rectifier bridge B1 converts this to a pulsating DC voltage, filtered by C1 and C2. R1 and D1 form a simple 9 v regulator circuit. C3 and C4 provide additional filtering and RF bypassing. The varistor,

V1, suppresses high voltage transients on the power line that might damage the circuit. This part is not essential to circuit operation and can be left out if the device will only see light duty, but it should not be left out if this project is going to be plugged in and left running continuously (as in an advertising display). If a circuit will have to operate under adverse conditions, such as thunderstorms, a varistor is a necessity. The pilot light can be omitted, if desired.

The transformer that we used is rated at 450 mA. If the load connected will draw close to this, you will need a larger transformer and bridge rectifier; however, if you are switching this much power, you might want to see if 120 v lamps can be used and interface an optocoupler and triac to it instead.

If this circuit is to be used in an automobile, you can alter the power supply by eliminating everything to the left of C1 (except the fuse and switch). You might want to insert an automotive line choke in the power lead to help kill electrical noise on the supply line. If this circuit will ever be used in an automobile when the engine is off, lowering R1 to 220 ohms (1/2 w) might be a good idea. The fuse should be raised in value, depending upon the load you are driving.

CONSTRUCTION DETAILS

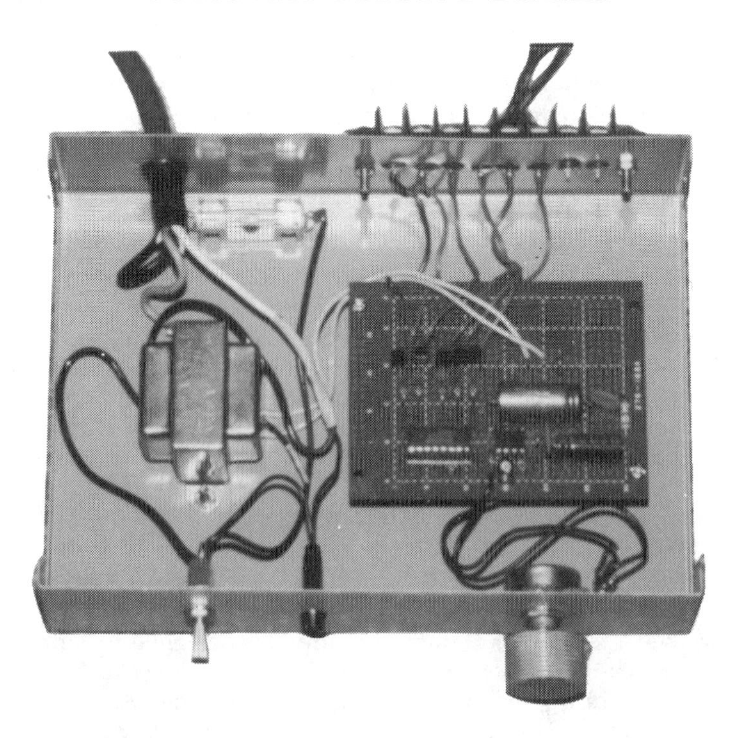

Interior view of a completed light sequencer project.

For this project, we used a metal enclosure from Radio Shack (#270-272A). This had more than sufficient room to house all of the components for this circuit. If you use this or any other metal enclosure, make certain to use a three-wire power cord and to ground the case.

We mounted the power switch, pilot light, and speed control on the front panel. If the speed of the sequencer will be permanently set, you can mount a trimmer potentiometer on the circuit board instead. To interface the light display to the circuit, we mounted an 8-position barrier strip on the rear of the cabinet, which has screw terminals to which leads from the light display can be attached.

The light display itself can take on many forms. LEDs, low voltage incandescent lamps, and 120 v AC bulbs can all be driven. LEDs are good for indoor displays and are very easy to drive with this circuit. Figure 7(a) shows how a single LED can be connected to each output. Often, you will want to drive many LEDs, such as in a light chaser display. You can connect several LEDs in series on each output. The value of the series resistor will drop as you add more LEDs (see figure 7(b)). Since the voltage drop across an LED is typically 1.5-1.8 v, only five or six can be put in this type of series, but you can add as many strings together in parallel as you want (see figure 7(c)). Each string of LEDs should have its own series resistor. With the 330 ohm resistor shown, each output should be able to handle at least five strings or thirty LEDs. Using five outputs would correspond to a display of up to one hundred fifty LEDs. More strings could be added in parallel if you are going to be sinking a lot of current, but it would be a good idea to replace the 2N4401 transistors with TIP120 Darlington transistors.

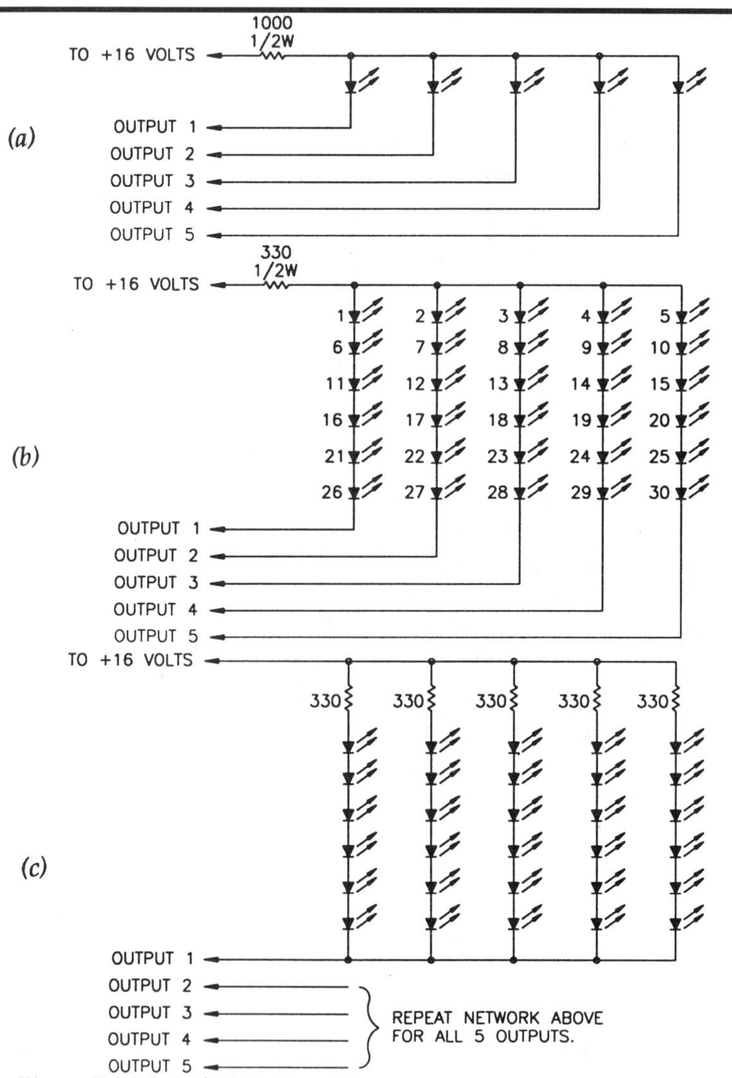

Figure 7 (a) Interfacing a single LED to each output; (b) increasing the number of LEDs by placing several in series. The number next to each LED gives the order to place them in a chaser display; (c) adding several strings in parallel on each output. If you place more than five or six strings on each output, change the output transistors to TIP120 Darlingtons. Don't forget, you must always use a current limiting resistor with LEDs, as shown in the diagrams above. The value of this resistor can be calculated by subtracting the total voltage drop across all of the LEDs (typically 1.5-1.8 v per LED) from the power supply voltage. This gives you the voltage drop across the resistor. Divide this value by the current flow you want through the LEDs (usually 10-20 mA.) in amperes, and the result is the resistance required. Don't forget to calculate the power dissipation required of this resistor (P=EI). A 1/4 w resistor may not be enough for the job.

Low voltage DC incandescent lamps can also be driven with this circuit. They tend to be "current hogs," however, and are not always the best choice, but in some applications, especially automotive ones, you may have no other choice. A sequential turn signal for a car is one example. Many of these bulbs will draw a considerable amount of current, and replacing the 2N4401 transistors with TIP120 Darlingtons is almost mandatory. Except for this detail, however, interface to the circuit is very simple. Figure 8 shows the connections.

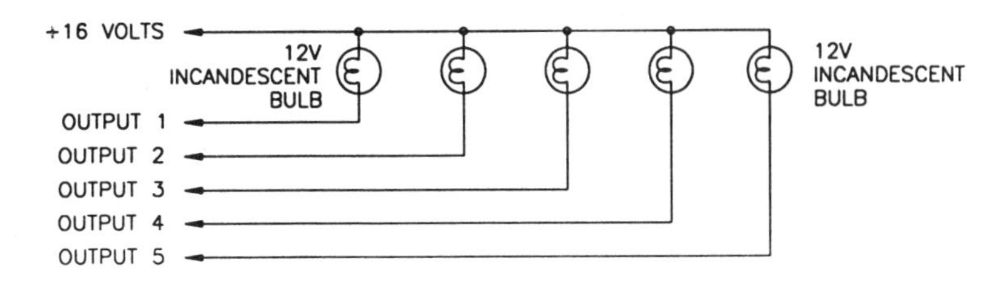

Figure 8 Driving 12 v incandescent bulbs.

Driving 120 v AC lights is useful for many applications, including advertising displays. Christmas lights are easy to attach because the lights are already wired in strings. Figure 9 shows how to interface AC lights to the circuit. The output transistor turns on the internal LED of a MOC3010 optocoupler. This, in turn, switches on the triac, which lights the lamp. Each triac can handle up to 6 amp and may need a heat sink, depending upon the load. The AC interface circuitry can be included in the project box, with AC outlets installed in the rear, or the AC circuitry can be enclosed in a separate box, interfaced with the control circuit through a cable.

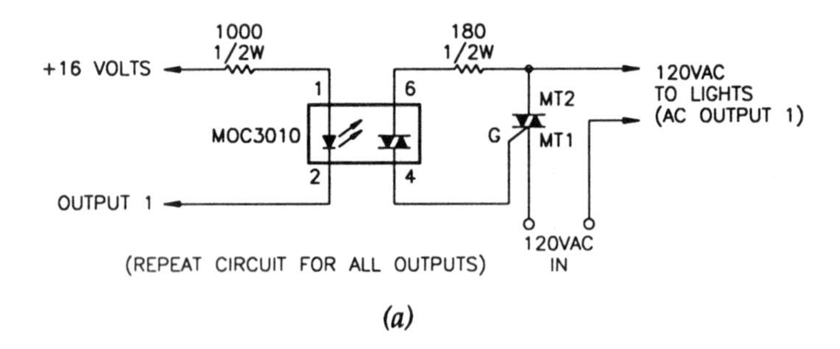

(a)

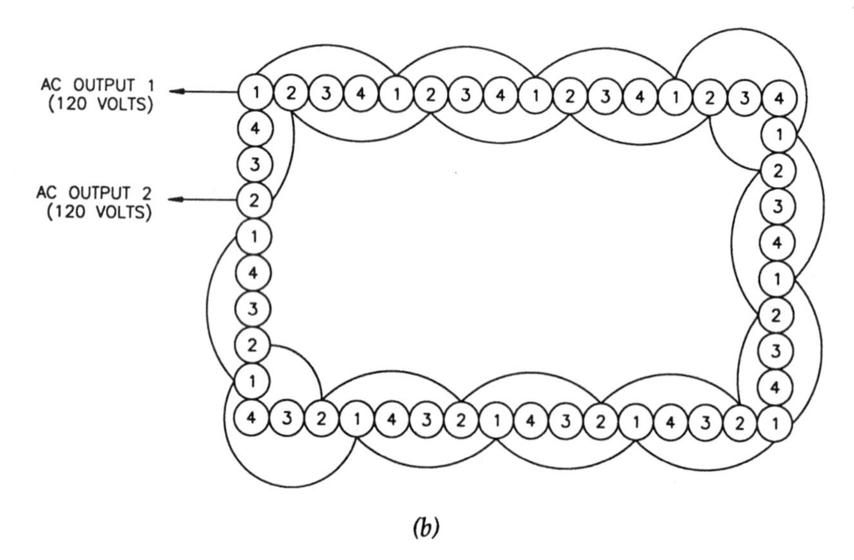

(b)

Figure 9 (a) Interface circuit for driving 120 v AC bulbs; (b) layout for a chaser display using strings of Christmas lights (120 v), one on each AC output. Using four strings of lights, as illustrated here, every fourth light should be on the same output. For clarity, we have shown the wiring for only the first two strings.

PARTS LIST

Semiconductors:

B1- 100 PIV-1.4 amp bridge rectifier (Radio Shack #276-1152)
D1- 1N4739-9.1 v-1 w zener diode (Radio Shack #276-562)
Q1, Q5- 2N4401 NPN silicon transistor (Radio Shack #276-2058)
U1- 555 timer IC (Radio Shack #276-1723)
U2- 4017B decade counter IC (Radio Shack #276-2417)

Capacitors:

C1- 1000 ufd-25 v electrolytic
C2, C4- .1 ufd-50 v disc
C3- 100 ufd-16 v electrolytic
C5- 4.7 ufd-16 v electrolytic or tantalum

Resistors: (All resistors 1/4 w unless stated otherwise)

R1- 390 ohm-1/2 w
R2, R6-R10- 4.7K-1/4 w
R3- 100K linear potentiometer (Radio Shack #271-092)
R4- 2.2K-1/4 w
R5- 10K-1/4 watt

Miscellaneous:

F1- 1/2 amp fuse and holder
S1- 120 v AC switch-amp
T1- 12.6 v-450 mA. transformer (Radio Shack #273-1365A)
V1- varistor (Radio Shack #276-570)
Circuit board (Radio Shack #276-168A)
8-position barrier strip (Radio Shack #274-653)
Pilot light (Radio Shack #272-712)
Project enclosure (we used a Radio Shack #270-272A metal box)
Knob for potentiometer, circuit board standoffs, line cord, misc. hardware

BRINGING UP THE UNIT

After connecting the display to the circuit and turning it on, the lights should come on and automatically start to sequence. Turning the potentiometer, R3, should cause the lights to slow down and speed up.

If the unit does not seem to function properly, check the following:

- If the lights do not come on at all, first check to make sure fuse F1 has not blown. If it is good, check all power supply voltages. There should be 15-18 v across C1 and close to 9 v across D1. If there is no voltage across C1, check the wiring of the transformer and bridge rectifier for any errors. If the voltage across C1 is correct, but the voltage across D1 is low, make certain D1 is not installed backwards and that there is no short across the 9 v supply lines to the ICs. Verify that the 9 v is reaching the 4017 IC by measuring across pins 16 (+) and 8 (-).
- If this checks out, disconnect the display and temporarily connect the positive lead of an LED to the +16 v supply. Jumper the LEDs cathode to one lead of a 1000 ohm resistor, and connect the other lead of the resistor to Output #1. The LED should blink; if so, the

problem is in your display. If the LED does not come on, check the wiring around the 4017 for errors. Check pin 14 to make certain that a clock signal is present (the voltage will go high and low repeatedly-at higher clock speeds, the voltage will average out to 5-7 v). If everything seems correct, replace the 4017 IC.

- If one string of lights comes on, but the lights do not sequence, this is the result of the 4017 not receiving a clock pulse. Check the voltage on pin 14 of the 4017 to ground. It should go high and low repeatedly. If the speed of the oscillator is turned up, the voltage will average out to read 5-7 v. If it is near 0 or 9 v, the clock stage is not functioning. Check all connections around the 555 IC carefully.
- If the 555 is outputting a clock pulse, but the 4017 still does not advance, make certain pin 15 (reset) is not accidentally shorted to pin 16, which would hold it in a reset condition. Make certain that pins 13 and 8 of the 4017 are solidly grounded. If everything seems correct, replace the 4017 IC.
- If the lights come on and sequence, but one string does not come on, measure the voltage on the pin of the 4017 that corresponds to that output. If it is going high and low, then the 4017 is good, and either the output transistor for that channel is defective, or else there is a problem in that leg of the display. Swapping wires to another output should quickly pinpoint which is at fault; however, if one output of the 4017 is not going high, the IC is probably bad. First, make certain that the output of the IC is not accidentally shorted to ground through a solder bridge. If the non-functioning output happens to be the last output, make sure the reset terminal (pin 15) is tied to the correct pin. It should be connected to the first unused output.

TOUCH-OPERATED SWITCH

This chapter describes the construction of a touch-activated switch. When properly assembled, all it takes is a touch of a finger to a metal plate or any small metal object to turn a device on or off. Depending upon the output configuration that you choose, the touch-activated switch can be used to simply latch a device on, turn it on for a specific period of time and automatically cut it off, or it can be used in a toggle mode, so that one touch will turn it on and the next will turn it off.

Good applications for these switches include a switch to turn on a bedside lamp (it is much easier to find a metal sensor plate than a lamp switch in the dark), alarms to protect valuables (attach the circuit to any small metal object and an alarm can sound if the object is touched), and alarms to protect children from dangerous items, such as guns (it can be attached to either a gun or the handle of a gun cabinet). These are also handy to turn lights and other devices on or off for the elderly and handicapped. If you are a real animal lover, you can even mount a sensor plate about one foot high beside your front door and wire the circuit in parallel with your doorbell button. Whenever your dog wants in or out, all he has to do is touch the plate with his nose to ring the doorbell.

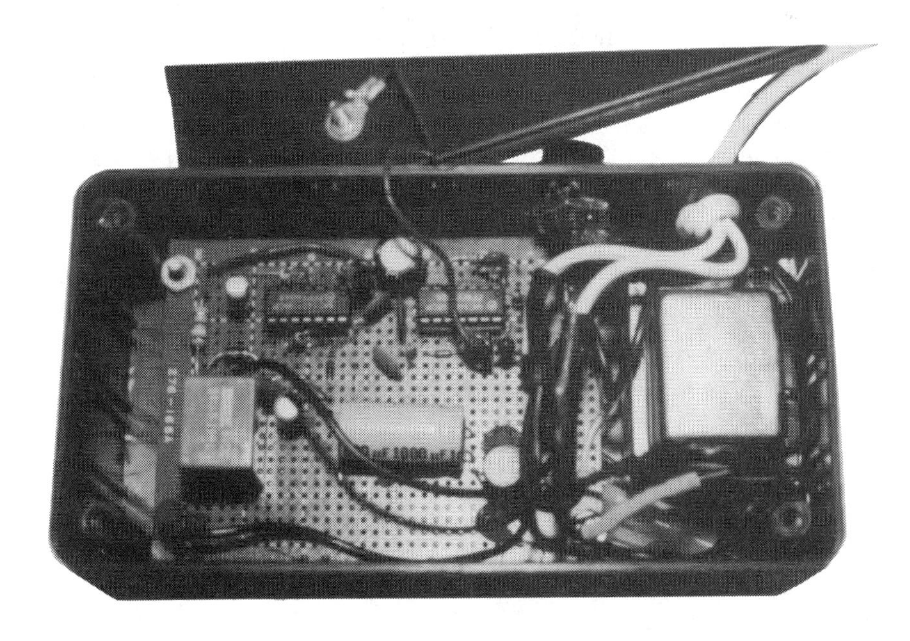

Interior view of a completed touch-operated switch project.

CIRCUIT DESCRIPTION

There are several approaches that can be taken to design a touch-activated switch, and all have their advantages and disadvantages. Touch-operated switches have a reputation for unreliable operation, and if not properly designed and installed, will live up to this reputation. Broadly categorized, there are three techniques commonly used for building these switches:

1) Resistive - the circuit is designed to detect a person's skin resistance.

2) 60 Hz Coupled - the human body acts as an antenna, picking up the 60 Hz power line signals surrounding us. An amplifier circuit with a high input impedance can be used to detect this.

3) Capacitive - the human body's capacitance to ground (either circuit or earth ground) is detected.

The first technique, the resistive, will not be discussed at any length because of its limitations. Basically, the idea is to use a sensor plate that has alternately connected lines across it. When a finger touches the plate, the skin resistance across two lines causes a small current flow, which can be amplified to operate a relay or other device. Although this will work, it requires a specific type of sensor and cannot be used, for instance, to detect that someone is touching the doorknob on your front door. This severely limits its usefulness for most applications.

The other two approaches are more practical. Of the two, the 60 Hz coupled-type is usually the easiest to build and the one most often used by electronics hobbyists. The detector for a 60 Hz coupled circuit is basically just an amplifier with a high input impedance. FETs and Darlington transistors were often used as input devices in the past, but ICs are more commonly used today. Figure 1 shows a typical circuit. With no one touching the plate, the input of the CMOS gate is held at ground, and the output stays high; when the plate is touched, the body acts as an antenna, coupling a 60 Hz signal to the plate. If the voltage is of sufficient amplitude, the output of the gate will switch, going high and low at a 60 Hz rate. The output signal can be used to latch a flip-flop or trigger a timer circuit.

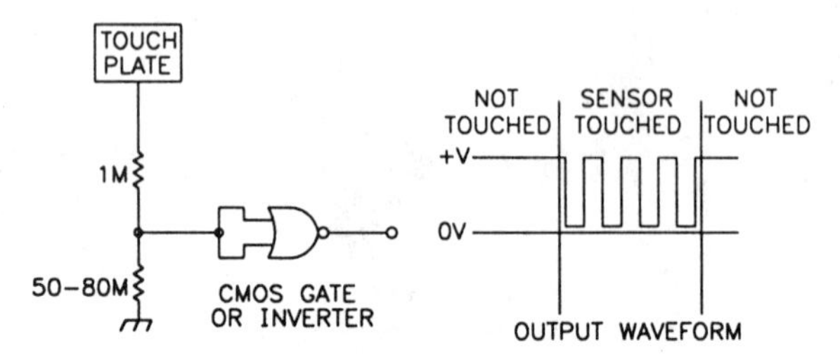

Figure 1 A 60 Hz coupled detector.

Some limitations do apply to this method. This circuit must be connected to an earth ground to operate correctly. The 60 Hz signal that your body is receiving is referenced to earth ground. Often, there is sufficient AC coupling to ground through the transformer in a power supply for this to work on power line operation without a direct connection, but stability is greatly enhanced if it is solidly tied to earth ground. If batteries are to be used, a line from circuit ground must be connected to earth ground for it to function at all. This does not necessarily mean you have to attach it to a ground rod; it can be connected through a power cord to the ground lead at a power outlet or connected to a cold water pipe. You should also be aware that this circuit will not work everywhere. Since it operates by picking up the 60 Hz signals surrounding you, you must be in a place where 60 Hz signals exist in at least moderate strength. With this concern and the earth ground hassle, this is not the circuit to use on your next camping trip.

The capacitance-operated switch is often the most practical. In many applications it will work equally well with or without an earth ground connection and works fine with batteries. It's less susceptible to false alarms than the 60 Hz coupled devices, and, since it does not depend upon an earth ground connection, it can be used in cars, campers, and RVs. Figure 2 shows one common configuration.

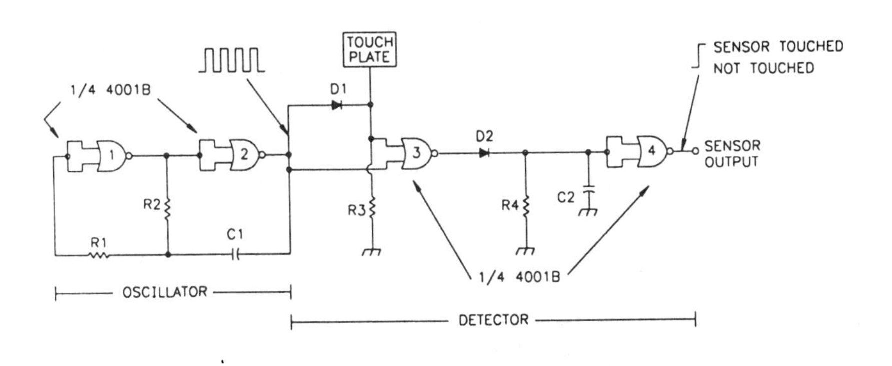

Figure 2 A capacitance-operated touch sensor circuit.

We start by building a high frequency oscillator. By generating our own squarewave signal, we are no longer dependent upon the reception of 60 Hz signals. The existence of a 60 Hz power field is no longer necessary, and neither is the earth ground connection to detect it.

The oscillator supplies a squarewave signal of about 40 kHz to both inputs of gate 3 in figure 2. When the touch plate is not being touched, both inputs of gate 3 follow the clock signal, so that the output of this gate is a steady squarewave signal. When the output is high, current flows through D2 and charges capacitor C2, holding the inputs to gate 4 high. Resistor R4 starts to discharge C2 on the half-cycle that D2 is not conducting, but as long as the squarewave signal exits gate 3, current through D2 will keep C2 charged enough to keep the output of gate 4 low.

When the plate is touched, the capacitance between the person touching the plate and the circuit ground, along with D1, causes a positive DC voltage to form on the input, holding it high (see figure 3). You might think of this as sort of a half-wave power supply, where the diode conducts on every positive half-cycle and the capacitor stores the charge to maintain the voltage level when the diode is off. As long as the capacitance is sufficient to hold the input high during the low half-cycle of each waveform, the output of gate 3 is held low, rather than following the clock signal, as before. If the plate is held long enough for C2 to discharge through R4, the output of gate 4 will go high, indicating that the plate has been touched.

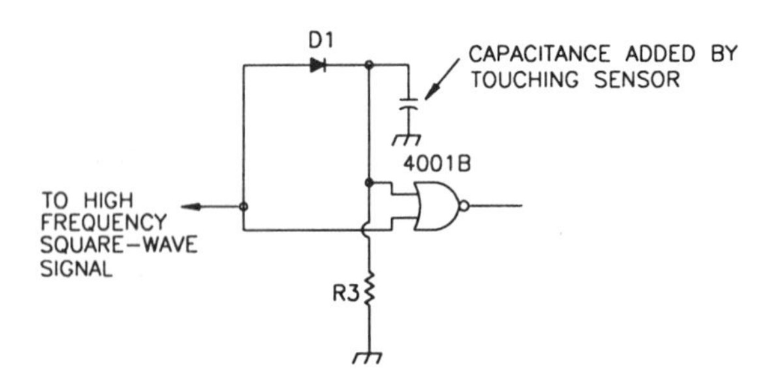

Figure 3 Equivalent circuit of capacitive detector with sensor plate being touched.

This technique also has its limitations. The main one is that while this system is not dependent upon an earth ground, the distance that the sensor can be located from the circuit is limited without it. Remember that this circuit operates by detecting a capacitance between the sensor connection and circuit ground (see figure 3). The body of the person touching the sensor can be thought of as one plate of a capacitor, and the wiring on the circuit board as the other plate. The closer the two plates of a capacitor are, all else being equal, the greater the capacitance; therefore, the sensitivity of this circuit drops off quickly if the touch plate is not close to the sensor board. For reliable operation with an ungrounded system, the sensor should be within 18" of the circuit. This is fine for a touch pad or alarmed display case where the circuit board can be enclosed underneath. In some applications where space is limited, the sensor board can be mounted near the sensor with wires run to it from the power supply and relay.

If the sensor must be more that 18" from the circuit board, you have the option of either using the 60 Hz coupled detector or simply grounding the capacitive switch. In either case, the circuit will require an earth ground. By connecting the capacitive switch to earth ground, we are using the earth as the other plate of our capacitor, instead of just the circuit ground. When the person touches the sensor, the capacitance from their body to the earth triggers the device; therefore, close proximity to the circuit board is no longer required, freeing us from the 18" limitation. The added wire attached to the input also has a certain amount of capacitance to ground, however, and the circuit sensitivity must be reduced to compensate for this.

Although the capacitive switch is often thought of as a switch for "close-range" operation, such as for musical keypads, it also works quite well at longer ranges with the sensitivity reduced and a proper ground. In our tests we could only attach about a 6-7 ft length of wire to the 60 Hz coupled detector shown in figure 5 without it going off by itself. Reducing the sensitivity on it any further would have prevented it from reliably activating when touched. On the other hand, by attaching an earth ground to the capacitive detector and reducing R3 to 220K, we were able to attach a 15 ft wire from the circuit to a doorknob. The circuit consistently operated without any sign of false triggering, although we would not encourage you to attach this long an input to any touch-operated switch. The 60 Hz coupled detector is also suitable for some short to mid-range applications; however, the higher input impedance makes it more susceptible to false alarms. It's also easier to reduce the sensitivity of the capacitive switch enough to prevent false triggers, while at the same time maintaining enough to set off the device when you touch it. This is especially true when a lot of metal is attached to the input, such as the body of a lamp. Our recommendation is to go with the capacitive switch, unless you feel that the 60 Hz coupled switch will fit into your application better.

We used the capacitive technique for our prototype. Figure 4 shows the entire touch sensor circuit used for controlling a lamp. One touch of the plate turns the lamp on, and the next touch turns it off. Notice the location on the schematic marked "point A." This point separates the actual touch detector from the output control circuit. Different output configurations can be attached at this point, allowing you to customize the design to your own particular needs. Also, the 60 Hz coupled detector in figure 5 can be mated to any of the output configurations at this point, if you wish to experiment with it.

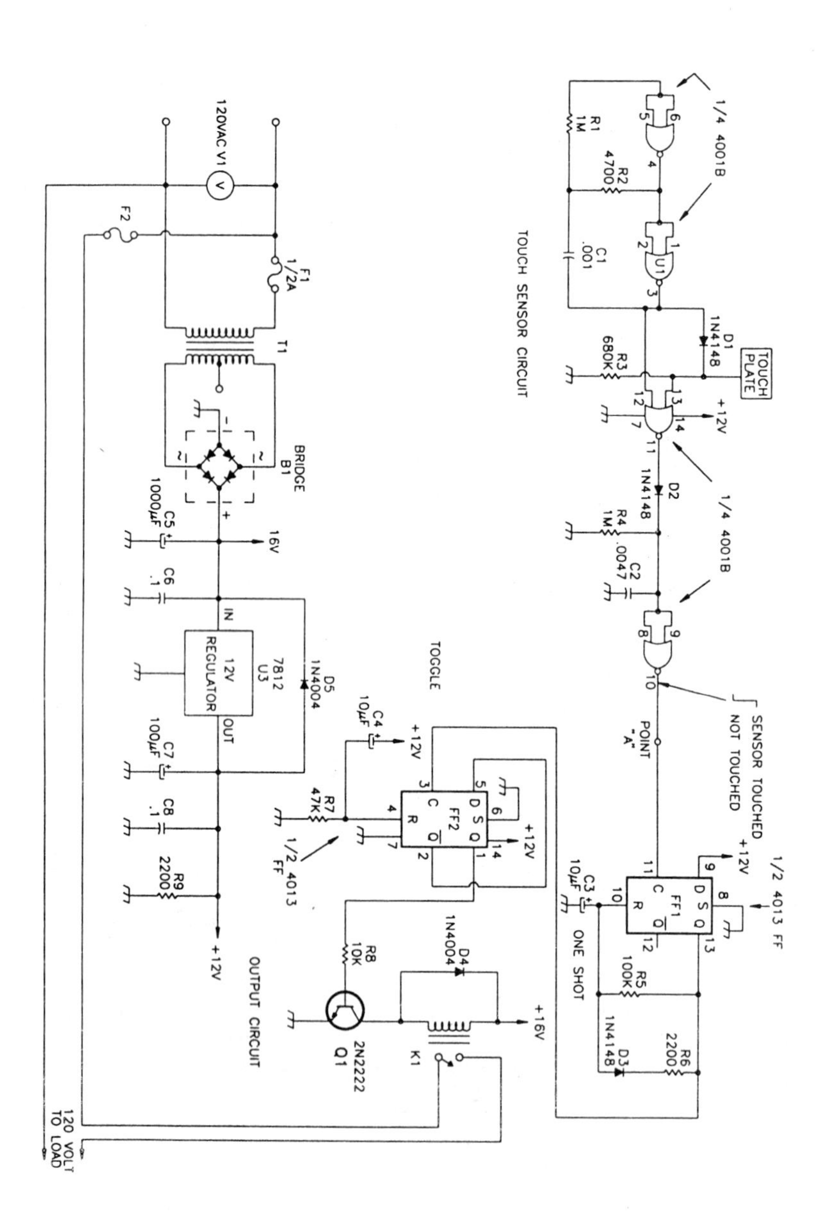

Figure 4 Complete schematic of the capacitance-operated switch.

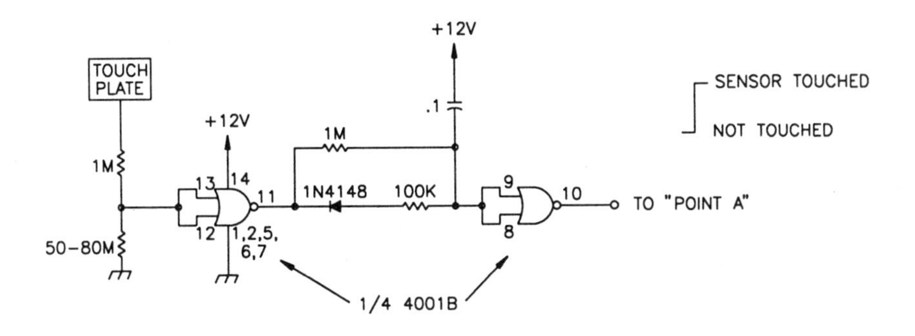

Figure 5 This 60 Hz coupled detector can be used instead of the capacitive detector if it seems to fit your application better. Simply eliminate everything to the left of "Point A" in figure 4 and substitute this. If you use this circuit, an earth ground connection is required for reliable operation. You can obtain the 50-80 meg needed on the input by stacking several 10 meg resistors in series. Fifty meg is usually sufficient, but the sensitivity can be increased by adding more resistors.

There are three types of outputs presented that can be attached to either detector at "point A" in the circuit of figure 4. The first one is the "toggle," and it is the one used in our prototype and shown again in figure 6.

This circuit uses a 4013B dual flip-flop IC. When the sensor plate is touched (regardless of the type of detector used), "point A" goes high. This positive transition on the clock terminal of the first flip-flop (FF1) causes the Q output to go high. This is because the D input is tied to +12 v, and the logic value on the D input is always transferred to the Q output on the positive edge of the clock pulse with this type of flip-flop, but when Q goes high, capacitor C1 (in figure 6) starts to charge through the 100K resistor. Since this capacitor is connected to the reset terminal of the flip-flop, a point will be reached where the capacitor will have enough voltage on it to cause a reset. Q will then go low, and the 2.2K resistor and diode will quickly discharge the capacitor to prepare for another cycle.

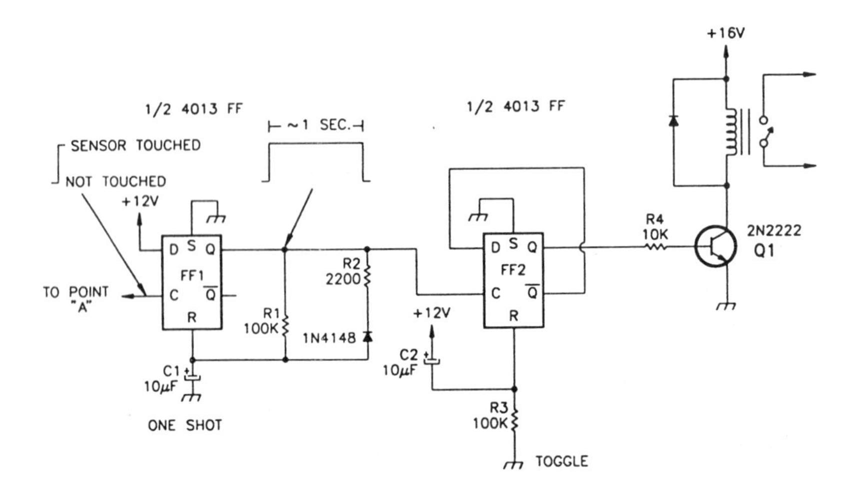

Figure 6 A "toggle" output stage.

In response to a touch of the sensor, FF1 will put out one long pulse. With the values shown, this pulse will last about two- thirds of a second. The reason that we want to do this is to prevent multiple triggers of FF2, which will do the actual toggling for us. It is possible that as your hand approaches or leaves the touch sensor, the circuit might momentarily switch on and off in hesitation. This first flip-flop guarantees that one, and only one, pulse will result. That is why this circuit, technically called a "monostable multivibrator," is often referred to as a "one-shot."

The second flip-flop (FF2), when its clock terminal goes high from the output of FF1, will alternate states. This is because the D terminal is connected to the inverted-Q output; therefore, when Q is high D will be low, so the next clock pulse will transfer this low value to Q; but as Q goes low, D will go high, so the following pulse will make Q go high again, and so on. The end result is that every odd pulse will make Q go high, and every even pulse will make Q go low. This is how we get our toggle action. The purpose of R3 and C2 in figure 6 is to momentarily hold the reset pin of FF2 high on power-up. This guarantees that the controlled device will be turned off when power is applied.

The second output configuration that can be used is the "latch." This output will turn on when the plate has been touched, and will not turn off until a button is pressed to manually reset it. This is useful for alarm circuits. Figure 7 shows this configuration. It is a very straightforward circuit, using only the "set" and "reset" terminals of one flip-flop.

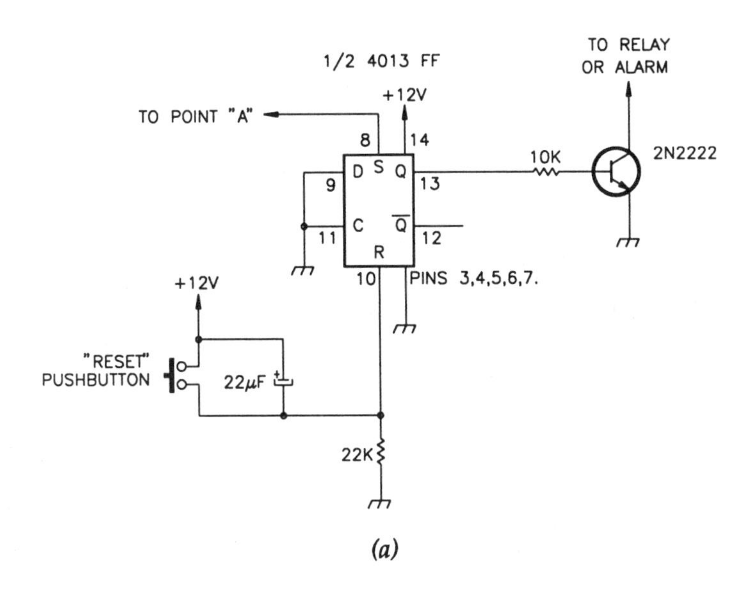

(a)

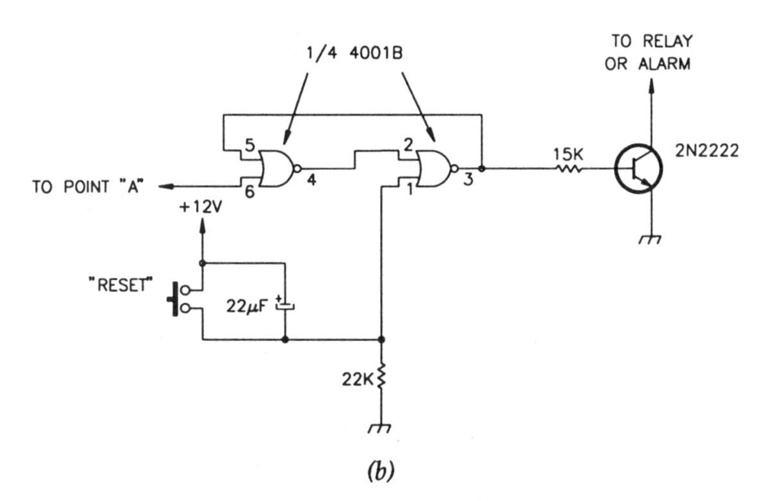

(b)

Figure 7 Two "latch" outputs. (a) A simple latch made from one-half of a 4013 dual flip-flop; (b) how to make a simple latch out of two NOR gates. This circuit can be used if you are using the 60 Hz coupled detector, since it only uses two of the four gates in the 4001B IC. With either circuit, the relay or alarm will energize when the plate is touched and will not turn off until the push-button is manually pressed. The 22 ufd capacitor in both circuits is included to guarantee that the load will be turned off on power-up.

If you are using the 60 Hz coupled-type detector, you can form a simple latch using the two remaining NOR gates and eliminate the 4013 IC.

The third output configuration is that of a "one-shot" (see figure 8). This is basically the same circuit as the first half of the toggle output. When the plate is touched, this circuit turns on for a predetermined amount of time and then automatically turns off. When we used it for the toggle output, the time period was set to a little under a second, but by changing the value of R1 or C1, we can make the delay as long as we wish, up to 10 minutes or so. This circuit is also a good choice for alarms, since it can sound for several minutes, and then reset itself. It is useful for some lighting applications, where you want to turn a light on and have it shut off after a delay. There are also some applications where a short delay is needed. If you are serious about letting your dog ring the doorbell, use a delay of about 3/4 of a second.

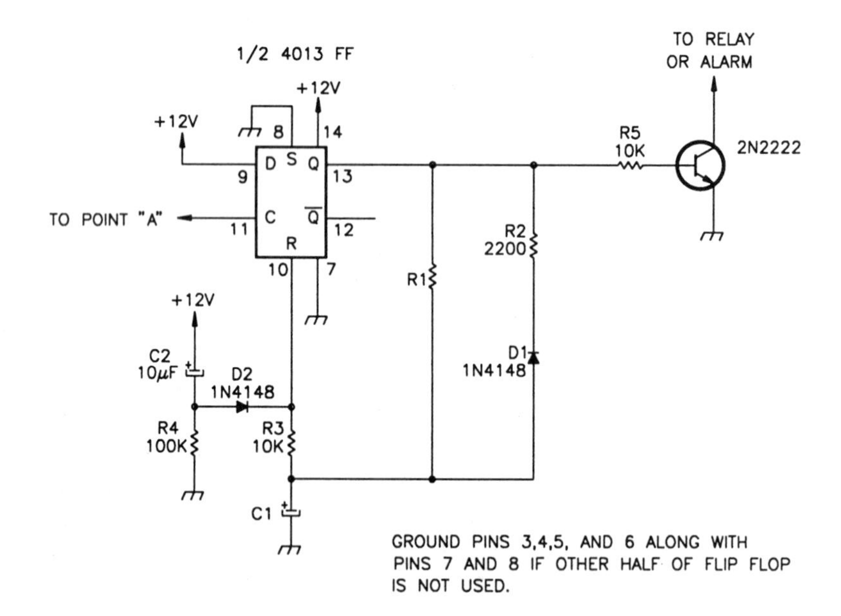

Figure 8 A "one-shot," or timer, output stage.

Since we explained how this circuit works while covering the toggle output, we will not go over the details again, but there are a few things we should mention. You may have noticed that the circuit is not quite identical. We have added C2, R3, R4, and D2. These components simply provide a reset pulse on power-up. You will also need to know how to determine the time delay. A close estimate of the delay can be calculated by simply multiplying the values of R1 (in ohms) and C1 (in farads) together, and then multiplying by two-thirds. You can also use units of megohms for R1 and microfarads for C1, which is often more convenient.

Suppose we wish to turn on a light for 20 seconds in a hall and then have it shut off automatically after we are through. A 1 meg resistor and a 33 microfarad capacitor would come close (a potentiometer can be used for R1 if necessary to adjust for exact delays). A 100K resistor for R1 and a 330 microfarad capacitor for C1 would also work. The accuracy of an analog delay such as this will not be as precise as that of a digital timer. On long time delays, there may be a slight variation in time from one cycle to another. This type of circuit is typically accurate to within a second or two per minute of delay. (If accuracy is much worse than this, try another capacitor for C1).

There is one other detail to watch here. For delays of more than a few seconds, it is usually necessary to use an electrolytic capacitor for C1 because of the large capacitance required. Electrolytic capacitors are notorious for leakage problems, and the larger the capacitor, the more likely it is to be a problem, especially 1000 microfarads and above. Figure 9 shows what can happen.

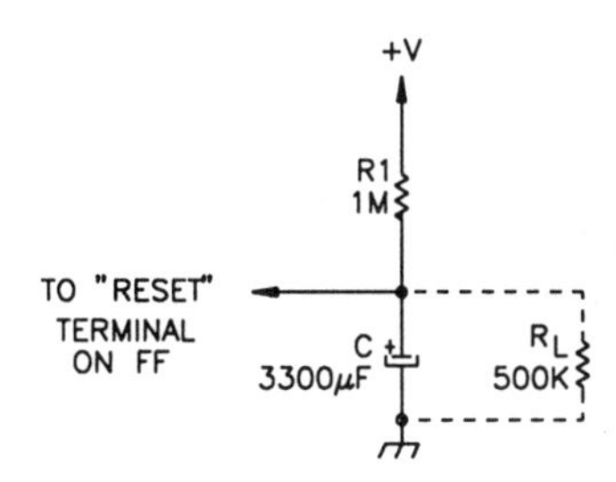

Figure 9 The leakage resistance of an electrolytic capacitor can form a voltage divider with R1, the timing resistor. If the leakage resistance is not high with respect to R1, the circuit may never reset. Attempts to achieve long timing durations with the popular 555 timer IC are likely to be frustrated by this same problem.

Suppose that you require a delay of about 40 minutes. You decide to try a 1 meg resistor for R1 and a 3300 microfarad capacitor for C1, which should be close enough for your application. Unfortunately, your capacitor has a leakage resistance of about 500 kilohms. When the flip-flop is set and C1 starts to charge, the voltage across it will never reach more than one-third of the supply voltage. When it reaches that point, the voltage divider set up by R1 and RL (the leakage resistance) halts any further increase; therefore, this circuit would never reset. The time delay would start, but never end.

If this circuit is to work, either R1 must be reduced considerably (shortening the delay), or C1 must be replaced with another capacitor. Not all will have the same leakage. You may find one that will work. However, designing a circuit that is going to require hand-picked parts to function is best avoided. This circuit is best for time delays of about 10 minutes or less.

Also keep in mind that the tolerance on electrolytics is very large. The actual capacitance may be almost twice what the device is marked at and still be in specification for some units. So if you build one of these with a 10 minute delay and it goes for 11 minutes without timing out, do not panic. The capacitance may just be higher than marked on the unit. Tantalum capacitors will give you better stability than regular electrolytics if you do not need more than 100 microfarads. Tantalums rated higher than this can be hard to find.

Regardless of the output configuration that you choose, the output of the 4013 flip-flop can supply only a small amount of current; therefore, we need to boost the current in order to control a relay or an alarm. The output from the flip-flop in our prototype (figure 4) drives a 2N2222 transistor, which, in turn, switches the relay on and off. For driving loads higher than 70 mA, you might want to substitute an IRF511 power MOSFET or a TIP120 Darlington power transistor for the 2N2222. These can switch much higher loads (at least 3 amp) without increasing the required drive current from the 4013 output. Although we show a relay being driven in the schematic, any suitable load capable of operating on 16 v, such as a buzzer or alarm, is fine. Most 12 v relays will operate at this voltage, but it does not hurt to insert a resistor in line to drop 3 or 4 v to keep it running cool. For switching 120 v AC loads, a simple solid-state relay, such as in figure 10, can also be used.

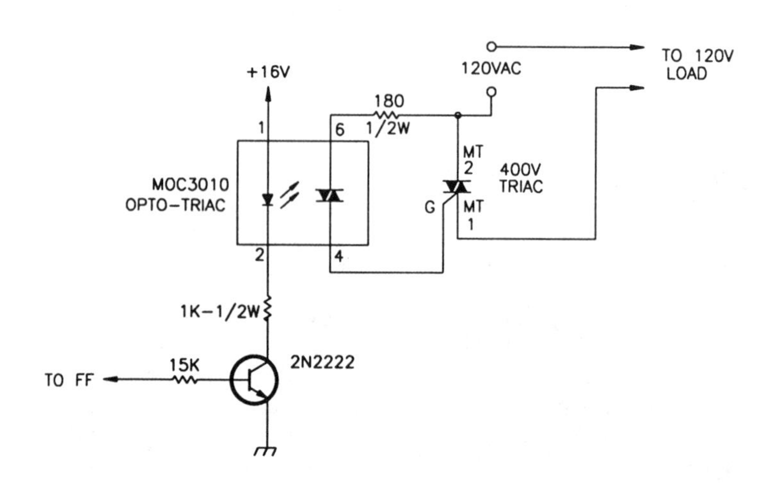

Figure 10 A simple solid-state relay suitable for turning on AC loads with the touch-operated switch. A heat sink should be attached to the triac unless the load is very light.

CONSTRUCTION DETAILS

The exact construction of this device will vary to a large extent, based upon the many different applications to which it is suited, but there are a few rules which will universally apply.

Our number one concern is safety. Every precaution must be taken to insure that the 120 v AC circuitry in the box does not come into contact with the control circuitry. A mistake could cause a serious shock hazard. Be especially careful on the circuit board itself, where the AC lines may attach to a relay. You may even want to mount any components at line voltage on a separate board. If the device will not be used to switch 120 v to a lamp or appliance, you may want to substitute a wall transformer, where high voltage will not need to enter the box. The whole idea is to make certain that when someone touches the sensor plate, they are coming in contact with an isolated, low voltage signal, and not 120 v. Wired correctly, this circuit is very safe, but extreme care must be taken to insure complete isolation from the AC line.

Another consideration is the housing for the device. We used a Radio Shack case (#270-223) to house the prototype. The transformer and a Radio Shack board (#276-168A) fit perfectly in this case. If you decide to use another housing, we recommend that you use a plastic case of some type. If the sensor plate will be mounted on the case, as ours was, we also recommend rubber feet on the bottom to prevent the box from sliding around. Some project cases will come with these. If the one that you select does not, you can get stick-on feet at Radio Shack and many hardware and electronics stores.

The circuit board that we used is a good choice for this project. If any 120 v lines are coming to the board, make certain that there are no solder bridges to adjacent contacts, particularly the +v or ground busses. Better yet, remove any adjacent copper areas to prevent this possibility. This can easily be done by holding your soldering iron on any unwanted copper area for several seconds. The bonding agent will release, and the copper can be easily removed. After verifying that the circuit worked properly, we also applied epoxy to the bottom of the board at all 120 v connections, just to be certain that they could not become free and shift position. Silicone rubber can also be used.

The touch plate can be any small metal object. Since the purpose of our prototype was to serve as a switch for a bedside lamp, we simply mounted a plate right on top of the box. The plate was actually the metal cover to a smaller Radio Shack case. A screw ran through the plate and cover, with a wire lug and nut attached underneath. The wire from the circuit input attached to the lug.

If the plate is going to be a little farther away from the circuit, one thing that may help guarantee reliable operation with the capacitive switch is to attach another wire to circuit ground, and run it in a loop around the sensor plate (see figure 11). The idea is to increase the capacitive coupling between the person in contact with the plate and circuit ground. By extending the ground in this manner, it may help to guarantee better reliability from all directions. If a plate is 18" away from the circuit, for instance, and no such ground loop is installed, the circuit would tend to work better if the person touching the plate is reaching over the circuit board, than if they are reaching from the opposite direction. A loop of wire surrounding the sensor and attached to ground can help overcome this effect. If the plate or object is on some type of platform, the ground loop can be mounted underneath and hidden from view. If you try this, do not get carried away running numerous turns on this loop. We are not building an inductor, only extending our ground plane.

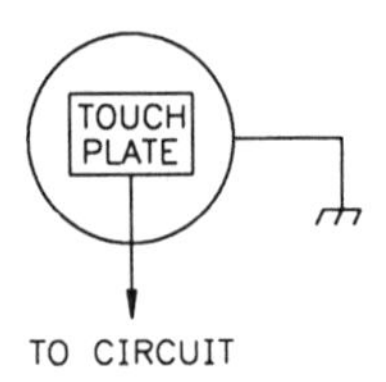

Figure 11 Running a wire ground loop around the sensor can help using the capacitive detector to insure reliable activation from all directions. If the touch plate is on a platform of some sort, such as might be used to protect an item on display from theft, the ground loop can be installed on the bottom side.

If you must mount the sensor more than 12" from the circuit, or the sensor is somewhat large, it is best to connect the device to earth ground, if at all possible. No ground loop will then be necessary, and the distance from the sensor to the circuit can be greatly increased; however, you will probably have to decrease the sensitivity of the circuit to prevent false activations or latch-up if you are running leads more than a couple of feet. This is done by changing the value of R3 (in figure 4). Lowering this value decreases the sensitivity of the circuit, allowing longer leads to be attached or larger amounts of metal to be used on the input (such as for a touch lamp control). With a 220K resistor for R3, input lengths of several feet are possible, while still maintaining stability. You may have to reduce R3 even more for some applications, but with any touch sensor circuit, the rule to remember is that the smaller the sensor object and the shorter the input lead, the more reliable the operation.

If you are using either type of detector with an earth ground, there are several ways to establish this ground connection. A separate ground rod can be used, or a connection to any copper cold water pipe is also acceptable, but the easiest solution is to simply use a three-wire line cord for the project and attach the ground (green) lead to circuit ground. As long as the power cord is plugged in, you will have an earth ground. If you are building this circuit for a burglar alarm application, however, keep in mind that the circuit may be disabled (especially with the 60 Hz detector) if the line cord is unplugged, even if you provide a battery back-up. One other detail to watch is that a "noisy" ground may cause more problems than it solves. If your circuit shows any tendency to turn itself on or off on occasion, make sure that the ground connection you are using is good.

With either type of detector, the smaller the amount of metal on the input, the better the operation. Keep your sensor relatively small and as close to the circuit as possible. If necessary, mount the sensor circuitry on a small board near the touch plate and run wires to it from the power supply and relay. If this is done, an additional .1 ufd disc capacitor should be connected across the 12 v power supply lines where they attach to the sensor board. This will help to suppress any noise induced on the lines coming from the power supply. The circuits that we have built (as in figure 4) have been extremely reliable. None of them have ever failed to operate or mysteriously turned themselves on or off, even through thunderstorms. There are not many commercial touch-operated products which can honestly make that claim.

PARTS LIST

Semiconductors:

B1- 100 PIV-1.4 amp bridge rectifier (Radio Shack #276-1152)
D1, D2, D3- 1N4148 diode (Radio Shack #276-1122)
D4, D5- 1N4004 silicon rectifier (Radio Shack #276-1103)
Q1- 2N2222A transistor (Radio Shack #276-2009)
U1- 4001B quad NOR gate - (Radio Shack #276-2401)
U2- 4013B dual flip-flop (Radio Shack #276-2413)
U3- 7812 12 v regulator IC (Radio Shack #276-1771)

Capacitors:

C1- .001 ufd-50 v disc
C2- .0047 ufd-50 v disc
C3, C4- 10 ufd-16 v electrolytic
C5- 1000 ufd-25 v electrolytic
C6, C8- .1 ufd-50 v disc
C7- 100 ufd-16 v electrolytic

Resistors: (All resistors 1/4 w unless stated otherwise)

R1, R4- 1 meg
R2- 4.7K
R3- 680K (a 470K and 220K in series can be used if you have difficulty locating this part)
R5- 100K
R6, R9- 2.2K
R7- 47K
R8- 10K

Miscellaneous:

F1- 1/2 amp fuse and fuse holder
F2- 2 amp fuse and fuse holder (a larger fuse can be used with a heavier relay, if the load requires it)
K1- 12 v power relay (we used a Radio Shack #275-005 9 v relay in series with a 390 ohm dropping resistor because of its low current consumption).
T1- 12.6 v-450 mA transformer (Radio Shack #273-1365)
V1- varistor (Radio Shack #276-570)
Circuit board (Radio Shack #276-168A)
Plastic case (we used a Radio Shack #270-223)
Circuit board standoffs, line cord, rubber feet for case, misc. hardware

BRINGING UP THE UNIT

Before applying power, use an ohmmeter to make certain that there is infinite resistance measured between circuit ground and either lead of the power plug. Then plug a load into the touch- operated switch and plug the circuit into an outlet. If the load you will be using for testing has a switch on it, make certain it is turned on.

When power is applied, the connected device should be off. Cautiously touch the sensor plate once. Obviously, you should not get shocked or feel a strong tingle. If you do, unplug the unit and find out what you did wrong before proceeding.

Upon touching the plate, the connected device should turn on. If you used the toggle output shown in figure 4, touching the sensor once more should turn the test load back off.

If the unit does not seem to function properly, check the following:

- Start by checking the power supply voltages. There should be at least 15 v across capacitor C5. If the voltage is low or zero, check the wiring of the transformer and the rectifier bridge, B1. Make sure that the fuse is good. If there is some voltage at this point, but it is well below 15 v, make certain that you are not using the center-tap connection on the transformer. Also, with the power off, measure the resistance across resistor R9 to make certain the 12 v power supply line has not been accidently shorted to ground through a solder bridge. If the voltage across C5 is satisfactory, measure the voltage across pins 7 and 14 of both the 4001B and 4013B ICs. There should be about 12 v. If there is at least 15 v going into the regulator, but the voltage coming out is wrong, make certain that the ground terminal of the regulator has a good connection and that there are no output shorts. If everything else seems correct, replace the regulator IC.

- If all power supply voltages are good, but the unit still does not work, touch the sensor pad while measuring the voltage on pin 10 of the 4001B IC. The voltage should be around 0 v before touching the pad and close to 12 v while touching it. If the voltage switches as it should, then the detector stage is working properly, and you can go to step 3 below, but if the voltage on pin 10 stays either high or low, the problem is in the detector. Assuming that you are using the capacitive detector, start by measuring the voltage on pin 12 of the 4001B. There should be 5- 7 v. If the voltage at this point is close to either ground or 12 v, then the oscillator is not working. Check the wiring of R1, R2, and C1 to the NOR gates for errors. If none can be found try replacing the IC. If this does not help, either R1, R2, or C1 is bad.

  If the oscillator is working, but the output is constantly staying low, short pin 13 of the 4001B to +12 v. If the output on pin 10 now goes high, check the touch plate input network of D1 and R3 for any errors. Make sure D1 is not bad. You may need an earth ground or ground loop around the sensor you are using. If shorting pin 13 to +12 v did not cause the output to go high, check the wiring of D2, R4, and the NOR gates. If these are correct, replace the IC.

  If the oscillator is working, but the output remains high continuously, the first thing to try is to disconnect the sensor from the input. If there is too much metal on the input for the sensitivity selected, the output will stay high. If removing the sensor makes the circuit function, reduce the value of R3 until the circuit operates reliably. If this does not help, then check D2, making sure it is good and not installed backwards. Then, if necessary, replace the IC.

- If the problem is in the output stage, a check of all wiring and voltage measurements on the 4013B IC pins should reveal the problem quickly enough. Particularly check the reset terminals to make certain that a capacitor with a lot of leakage is not holding them active. Neither reset terminal on the 4013B (pins 4 and 10) should have more than 1 v on it after power-up. If you are using the toggle output, you can divide this circuit in half by measuring the voltage on pin 13 of the 4013B IC. The voltage normally should read about 0 v, going high (close to 12 v) for almost a second whenever the sensor is touched. If the detector stage is working properly, but the output of the one-shot on pin 13 does not respond, the problem is in the one-shot circuit. If the output of the one-shot does go high, but the output still does not switch on, the problem is either in the toggle stage or the relay driver circuit. A voltage measurement on pin 1 of the flip-flop should reveal which is at fault. If the toggle stage is at fault, but everything else seems to be in order, replace the 4013B IC.

- Regardless of the output configuration that you are using, if the output going to R8 switches high, but the relay will not energize, the problem is in the relay driver circuit. Short the collector of Q1 to ground. If the relay energizes, and the transistor and R8 are wired properly, then the transistor is apparently bad. If the relay does not energize, the +16 v are not getting to the relay or else D4 is shorted or installed backwards.

- Once completed, if the circuit ever shows any tendency to mysteriously turn itself on or off (though none of ours have ever done this), there are a few things you can try to remedy this. The first thing to try is to reduce the sensitivity, particularly if you are using a rather large sensor. Reduce R3 as necessary to guarantee stability, even down to 100K or so. If you are using an earth ground connection, make sure it is not "noisy." This is most likely to happen if you are getting your earth ground through a power outlet. You might try connection to a cold water pipe or ground rod instead to see if this helps. Do not leave out varistor V1. It helps to suppress any high voltage transients coming in over the power line. If you can not find a convenient place to mount it in the box, you can use a spike suppressor on the power outlet instead, although this will be more expensive. You can also try to add additional .1 ufd disc capacitors across the 12 v supply lines on the circuit board, particularly near the 4013B IC.

TOUCH-CONTROLLED LIGHT DIMMER

This project is a touch-controlled light dimmer circuit. By touching a plate or metal object attached to the input, it will first turn a light on to a low level. A second touch will result in a 50% light intensity. One more touch will give full brightness, and a fourth touch will turn the lamp off again. This circuit is suitable for any incandescent lamp up to 240 w. Higher wattages can be driven with slight modification to the device.

This device is also available commercially; many places sell touch dimmers. They can probably be purchased for the same cost, or possibly even less, than what it costs to build this circuit, but there are still several reasons why this makes a good construction project:

- Reliability - Most of the plug-in dimmers commercially available do work, but they are easily fooled by noise spikes, brownouts, thunderstorms, etc. Almost anyone who has had a touch-controlled lamp has seen it turn on or off by itself during a storm. Although this is generally not an everyday occurrence, it is still enough of a problem to be a nuisance. No one wants to come home after a two week vacation and find half of the lights in their house turned on.
- Safety - This unit includes a transformer and optocouplers to keep the control circuitry totally isolated from the AC line. You may have wondered how manufacturers managed to get a transformer into the small package that their devices come in. The answer is that there is no transformer; these products almost always run "hot chassis." Although the techniques that they use can be safe when properly done (we have never heard of anyone being electrocuted by a touch lamp), such techniques are not for us. In all fairness, some manufacturers use a proximity effect in order to avoid any electrical contact with the circuit at all, especially in the "plug-in" type touch dimmers. These are indeed safe, but are probably the worst offenders in the reliability department.
- Versatility - By building your own unit, you can customize the design to suit your own particular needs. For instance, most commercial units either operate by touching a plate on a wall switch, or by touching the lamp itself. With this unit, the object that is touched does not have to be connected in any way with the AC circuitry of the controlled device. The sensitivity can be adjusted for optimum operation with the sensor used. Also, the number of "steps" or different lighting levels can be changed, as well as the intensity of the light at certain levels. We used three levels of lighting in our prototype, going from dim to medium to full on. Nine different levels can be selected, if desired, or perhaps you only want to use two. The choice is yours. You can also change the order of the light intensities. Normally, the light starts out

dim and steps up to full brightness, but you can reverse the order, if so desired, or have the light step up to full and then step back down to off. Although commercial units are designed to be generic, you can modify a unit that you build yourself to do whatever you want.

- Educational value - One of the advantages of building a circuit like this is that you learn how it operates and what makes it function. This is often the most important consideration, particularly for the young person considering a career in electronics.
- Accomplishment - There is almost always a feeling of accomplishment associated with having successfully built a working, useful device from scratch. Anyone can buy a commercial unit and plug it in, but it takes a certain degree of knowledge and ability to build one of your own.

If you want to build a touch dimmer, make certain that you also read the chapter on the 'DC-controlled Light Dimmer' because that project can also be used for this purpose. The DC-controlled dimmer may give slightly better response at the very dimmest levels. It is also easier to add additional lighting levels to that circuit because each new level only requires one diode and one resistor; however, its operation is slightly more complicated because it uses a different technique to generate the various light intensities. The dimmer presented in this chapter should be more than satisfactory for the vast majority of applications, but it may be worth your time to also look the other circuit over, if for no other reason than to demonstrate that there are often many different approaches that can be taken to solve a problem with electronics.

Exterior view of a completed touch-controlled light dimmer project.

CIRCUIT DESCRIPTION

Before starting this project, it is essential that you read the chapter on the 'Touch-operated Switch.' The sensor portion of that circuit is identical to the one used here, and its operation and troubleshooting are explained in detail. Once you understand the operation of the touch sensor circuit, we can proceed to examine the schematic of our dimmer.

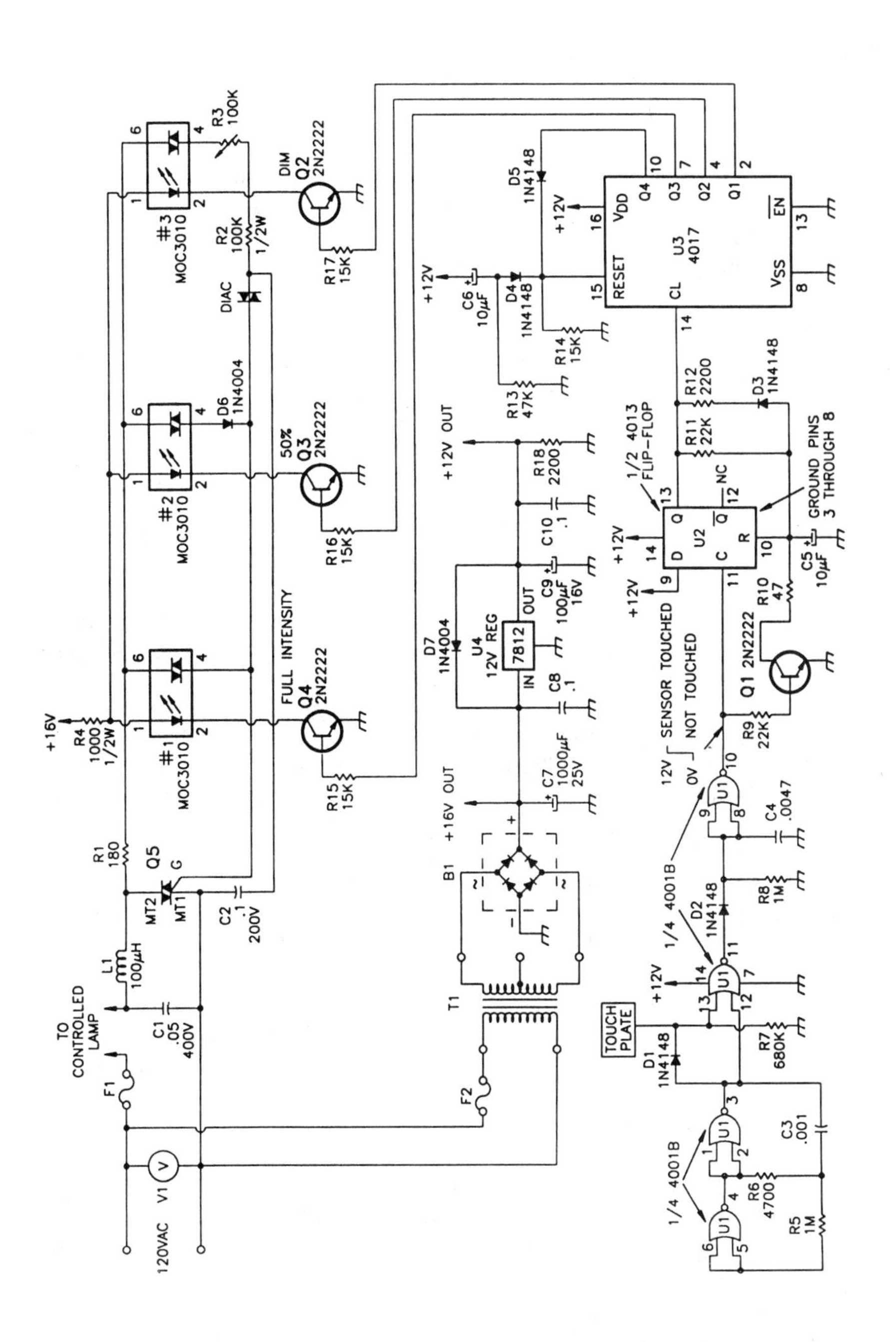

Figure 1 Complete schematic for a touch dimmer.

Figure 1 shows the complete schematic diagram of the touch dimmer. This circuit can be broken down into four major sub-systems: the AC power circuitry, the touch sensor circuitry, the stepper control circuitry, and the power supply.

The AC power circuitry consists of triac Q5, C1, L1, the three MOC3010 optocouplers, R1, R2, R3, C2, and the diac. These components are responsible for actually turning on the light when they receive a DC control signal from the stepper circuit. To see how this is accomplished, let's take a close look at one of the optocouplers.

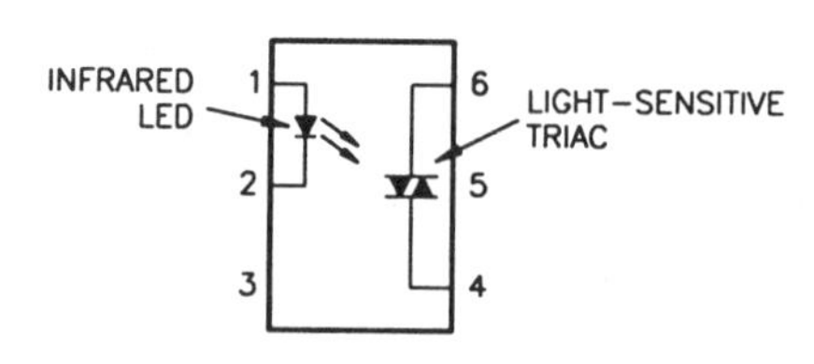

Figure 2 Internal components of the MOC3010 optocoupler.

Looking at figure 2, we see that the MOC3010 optotriac has two components: an LED (infrared) and a light-sensitive triac. Normally, with the LED off, the triac is in a high impedance state, similar to an open switch (although there may be a small amount of leakage); however, when the LED is turned on by a flow of current, the IR light that it emits causes the light-sensitive triac to switch to a low impedance state, similar to a closed switch. As long as the LED is on, therefore, the triac is also turned on. When the LED is turned off, the triac will continue to conduct until the current flow through it drops below a minimum level, known as the "holding current." This is a typical characteristic of all SCRs and triacs. With an AC signal applied, as in our application, the current flow will cease at the next zero-crossing of the AC line.

The reason that we are using these optocouplers is because they give us complete isolation between the DC control circuitry and the AC drive circuitry. Since the touch sensor is referenced to DC circuit ground, we do not want our DC circuitry at AC line potential. The optocoupler overcomes this problem by allowing the DC circuit to control the AC circuitry via transmission of light, rather than through direct electrical connection. Establishing a current flow through the internal LED allows our DC circuit to turn on the AC circuit, and stopping the current flow through the LED allows us to cut off the AC drive circuitry.

The triac in the MOC3010 is too small to handle large amounts of power, so we use this triac to trigger a larger triac, Q5 (see figure 1). If all we wanted to do was to turn the light on and off, we would only need one optocoupler, but since we want three different levels of light, we use three optocouplers, each connecting a different triggering network to triac Q5. The AC drive circuit can be more easily understood if you keep in mind that only one optocoupler is ever on at any given time and that the other two couplers, along with their associated triggering networks, are effectively removed from the circuit.

Figure 3(a) gives us an equivalent circuit of what triac Q5 sees on its gate terminal with MOC3010 #1 turned on. Since optocouplers #2 and #3 are off, D6, R2, R3, C2, and the diac have absolutely no effect on the circuit. But optocoupler #1 is turned on, allowing current flow through R1 to the triac gate terminal. The triac is, therefore, turned on fully, which in turn allows nearly full current flow through the lamp connected to the dimmer. This is how we obtain the 100% brightness level.

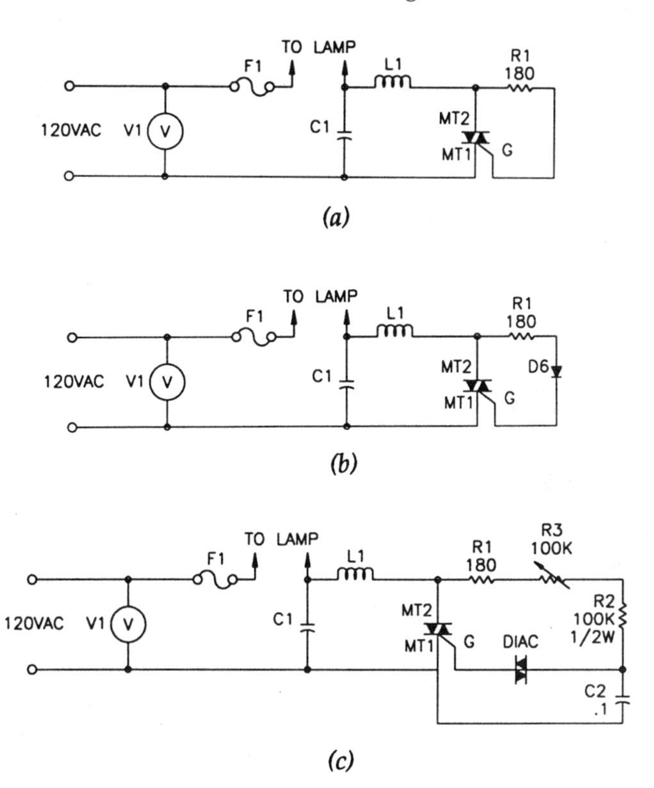

Figure 3 (a) Equivalent circuit of the AC driver network with optocoupler #1 turned on; (b) AC driver circuit with coupler #2 turned on; and (c) AC driver circuit with coupler #3 turned on.

When optocoupler #2 is on, the triac sees the network in figure 3(b) connected to its gate terminal. Once again, since couplers #1 and #3 are off, their associated components have no effect on the circuit operation. The network on coupler #2 is identical to that on #1, with the exception that diode D6 has been inserted. Because of this diode, triac Q5 will turn on every positive half-cycle but will remain off every negative half-cycle; therefore, the lamp attached to the circuit will be getting full current applied every half-cycle, or 50% of the time. This gives us our 50% light level. Diode D6 cannot be put in backwards; if you reverse its direction, the circuit will still work the same. The only difference is that the triac will conduct on the negative, rather than positive, half-cycle.

When optocoupler #3 is turned on, the triac sees the network in figure 3(c) connected to its gate terminal. This circuit is essentially a standard light dimmer. When the voltage on either half-cycle starts to rise, capacitor C2 starts to charge through R1, R2, and R3. The triac is not triggered, however, until C2 charges to the "break-over" voltage of the diac, usually 30-40 v. When this voltage is reached, the diac will partially discharge C2 through triac Q5's gate terminal, turning it on for the remainder of that cycle. Since the triac will always turn off at the next zero-crossing of the AC line, the portion of the cycle that the triac is on is determined solely by how fast C2 reaches the break-over voltage. If R3 is increased in value, C2 will charge slower, causing the triac to turn on later in each half-cycle. This will cause the lamp to dim. If R3 is decreased, C2 will charge faster, triggering Q5 earlier in the cycle. This will cause the lamp's intensity to increase; therefore, R3 can be used to set the precise light intensity desired for the dimmest level.

The second light level (the 50% level) can also be made adjustable, if desired. Inserting diode D6 gave us a simple, adjustment-free method of turning the lamp on at a 50% power level, but if you want this level to also be adjustable, simply replace D6 with a 100K trimmer potentiometer and move the lower connection to the other side of the diac.

If none of the optocouplers are turned on, the gate of the triac is essentially unconnected, and the triac is not triggered at all. The result is that the lamp remains off. We, therefore, have established a means by which the DC circuit can control the lamp connected to the dimmer, while remaining totally isolated from it electrically. Turning on the LED in optocoupler #1 gives us full lamp intensity; turning on the LED in #2 gives us a 50% power level; turning on the LED in #3 gives us a dim level whose intensity is determined by the setting of R3; and leaving all three optocouplers off turns the lamp off.

The DC control circuitry is composed of the touch sensor circuit and the stepper circuit. The touch sensor circuit is built around the 4001B IC. Its operation is described in detail in the chapter on the 'Touch-operated Switch,' so we will not elaborate on it. Basically, when the sensor plate is touched, the output of this circuit (pin 10 of the 4001B) will go high and remain high until the plate is released. Its function is to generate a pulse whenever the plate is touched, causing the stepper circuitry to advance to the next "step," or light level.

We do encounter one problem when we attempt to interface the touch sensor to the stepper. If our touch sensor hesitates when our hand first makes contact with the sensor plate, or as our hand is removed, it will generate spurious pulses in addition to the intended one. The stepper IC may see this as multiple inputs and advance more than just one step. To get around this problem, we interface the stepper and touch sensor circuit through a monostable multivibrator, or "one-shot." This circuit "debounces" our signal and guarantees that one, and only one, pulse is generated for each touch of the sensor plate.

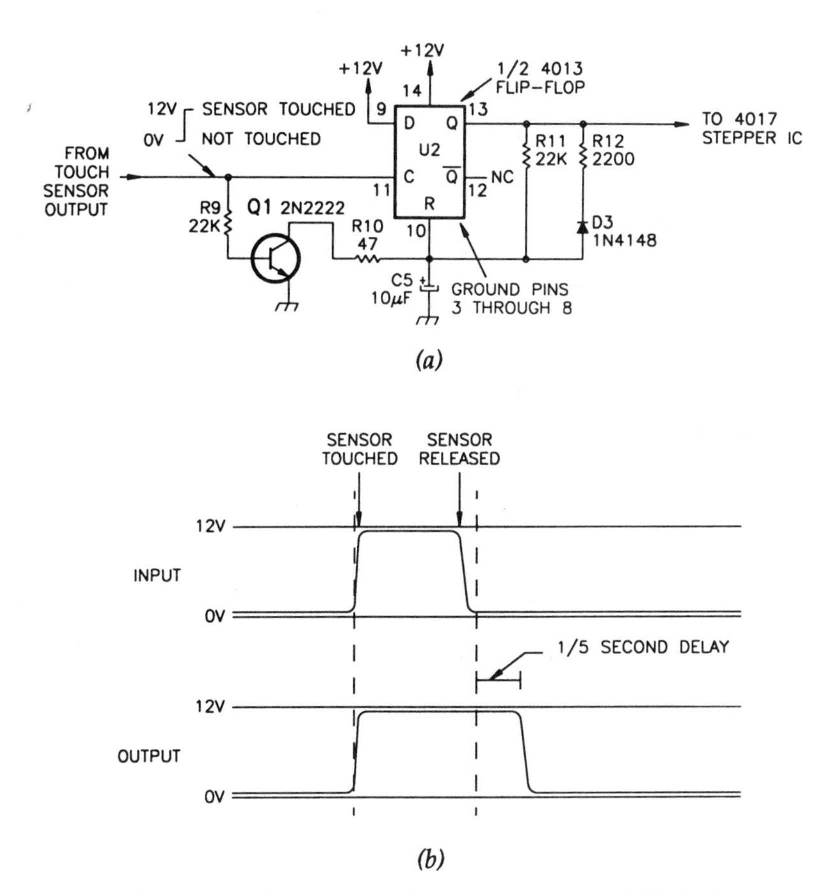

Figure 4 (a) Schematic of the one-shot stage; and (b) timing relationship between the input and output of the one-shot.

Figure 4 shows the one-shot circuit. It is built around one-half of a 4013 D flip-flop IC. In the 'Touch-operated Switch' project, we used a one-shot circuit to generate a pulse of nearly 1 second duration. This, in turn, triggered a flip-flop connected in a toggle mode. When the plate was touched, the output of the one-shot went high, triggering the flip-flop. It then remained high for almost a second, giving the person time to release the sensor before going low; therefore, it generated one pulse per touch.

Although that technique is fine for a touch switch, it would prove cumbersome for a touch dimmer. There will be times when the sensor will be touched three times in rapid succession, such as when it is desired to go from full off to full on. Having to wait over a second between touches would be frustrating, but shortening the delay too much will allow multiple triggers. We get around this dilemma by shortening the delay (to about 1/5 second) and making the one-shot re-triggerable. Looking at figure 4, we see that when the touch sensor output first goes high, the clock terminal of the 4013 one-shot is triggered. This causes the Q terminal, which clocks the 4017 stepper IC, to go high. C5 would normally charge through R11 and cause a reset of the 4013 in under 1/5 second, but as long as the clock terminal of the 4013 is high, Q1 is turned on, holding C5 in a discharged state; therefore, the time delay will not begin until the clock terminal goes low. If it goes low momentarily as the sensor is first touched, but goes high again after firm contact is made with the sensor, any charge accumulated on C5 is again discharged. The end result is that once this circuit has been triggered, its output will not go low again until the sensor plate has been released for a full 1/5 second. This allows us to touch the lamp or sensor plate quite rapidly (3 times per second), while still generating only one pulse per touch.

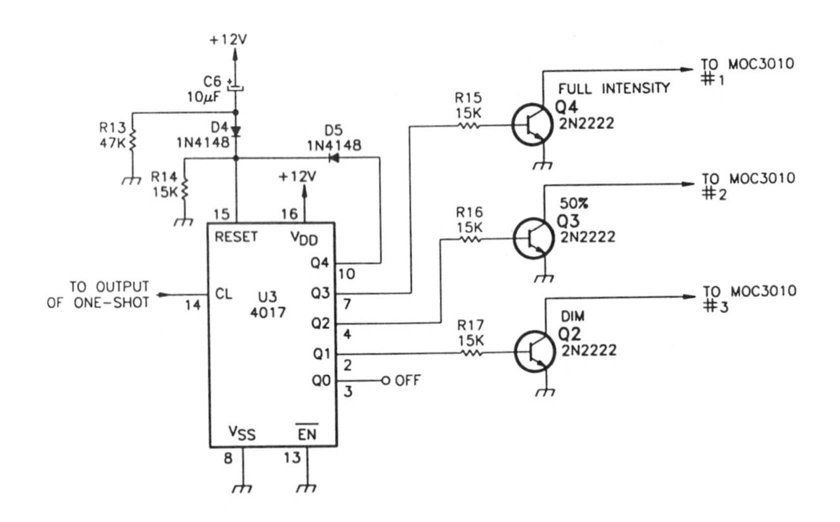

Figure 5 Schematic of the stepper circuit.

Looking at figure 1, we see that the output of the one-shot is connected to the clock terminal of the 4017B IC. This chip is a decade counter with a 1-of-10 output code. Unlike some decade counters, such as the 7490, which have a binary output, this counter has a separate output for each of its ten possible states. The first output, output 0 (called Q0), is high when the internal count is zero. The other nine outputs will be low. When a clock pulse is received, the counter will increment on the low to high transition of the signal. Output 0 will then go low, and output 1 (Q1) will go high. The next clock pulse will cause Q1 to go low and output Q2 to go high, and so on. When output Q9 is high and a pulse is received, the counter resets to zero and starts over. If all ten output states are not needed, the count length can be shortened by tying the first unused output to the reset terminal.

Figure 5 shows the stepper stage used in this project. We use the 4017 to increment our light intensity levels. On power-up, C6 and R13 provide a momentary reset pulse through D4, clearing the counter to zero. As a result, output Q0 will be high after power is applied. This output is left unconnected, because at this point we want the controlled lamp to be off. When the sensor plate is touched, the one-shot is triggered, and it clocks the 4017 with one pulse. This causes output Q0 to go low and Q1 to go high, turning on transistor Q2. When this transistor conducts, the LED in MOC3010 #3 is turned on. As described earlier in our discussion of the AC circuitry, this will cause the attached lamp to glow at a dim level. When the sensor is touched again, another clock pulse is applied to the 4017. This causes output Q1 to go low and output Q2 to go high. Now transistor Q3 will be on, which will turn on the controlled lamp to a 50% brightness level. One more clock pulse causes output Q3 to go high, and transistor Q4 is turned on, causing our lamp to run at full intensity.

If another pulse is received at this point, output Q3 will go low, and Q4 will go high. We could have another light intensity level here, if we desired, by simply inserting another optocoupler and transistor driver circuit. In fact, we could have nine levels of intensity, since we have ten outputs, but three intensity levels is a good number for this application, so we coupled output Q4 back to the reset terminal through D5, causing a reset whenever Q4 is high. Once reset, the counter is set to zero, Q0 goes high, and the lamp is turned off.

If you do want to add some additional intensity levels, simply insert another optocoupler and transistor driver circuit identical to the three shown in figure 1. The only difference between these three is the network attached to pin 4 of the optocoupler. For any additional intensity levels, you can connect a 100K potentiometer from pin 4 of the new MOC3010 to the point in the circuit where C2 is connected to the diac. The potentiometer can be used to adjust the intensity.

The power supply is composed of T1, bridge B1, C7, C8, C9, C10, D7, R18, and the 12 v regulator. The transformer applies 12.6 vac to the bridge, which rectifies the signal. C7 acts as a filter to smooth the ripple, while C8 serves as an RF bypass to help filter any transients. The regulator outputs a steady 12 v to C9 and C10, which help with transient response and, once again, provide bypassing. D7 does not perform a function in normal operation; its sole function is to protect the regulator in the event that the voltage on the regulator's input drops faster than that on its output during power-down, or if the voltage on the input is accidently shorted to ground. Although neither of these possibilities are very likely here, a diode is cheap insurance. The purpose of R18 is to draw about 5 mA from the power supply continuously to help prevent the problem just discussed, and to maintain a required minimum current flow through the regulator.

CONSTRUCTION DETAILS

Interior view of a completed touch-controlled light dimmer project.

Because a capacitance sensitive switch is part of the circuitry in this project and much of the circuit is operating at line voltage, we recommend that a plastic case be used to enclose the touch dimmer. We used a Radio Shack case (#270-224) for the prototype, which had more than sufficient room for all the components. We put all of the low voltage components on one board (a Radio Shack #276-168A), and most of the 120 v components on another (a Radio Shack #276-159) in an attempt to make certain that the line potential could not accidently be shorted to the DC circuitry. It is very important to make certain that the AC driver circuit and the DC control circuit are totally isolated to prevent a shock hazard. Using an ohmmeter, you should measure an infinite resistance from either terminal of the line cord to the DC circuit ground.

You should have no trouble finding most of the components for this project; all but one can be found at Radio Shack. The one exception is the diac. Although Radio Shack will not have this, you can obtain one from many of the mail order electronics suppliers, or from most electronic wholesalers. We used the ECG 6408 in our circuit, primarily because most of the wholesalers, and even many TV repair shops, either stock or will order ECG parts, or some equivalent. If you obtain a diac from another source, feel free to substitute another part number. Just about any diac will work, but if you have a choice, select one with a break-over voltage around 32 v.

We have rated the maximum load for this project at 240 w. This is primarily because of L1, which has a 2 amp rating. This inductor, along with C1, serves as a filter to suppress any RF interference generated by the triac switching on. If you must control a larger load, this inductor must be replaced with one having a larger current capacity. For loads under 400 w, you may be able to use two of the chokes listed in parallel. This will reduce the effectiveness of the filter, but double the current rating. For loads much greater than this, you will either need a larger inductor, or else you will have to take it entirely out of the circuit, shorting across its connections with a wire of suitable current capacity. This is not essential to circuit operation; in some circumstances the RFI may not be a problem. If you do power loads over 200 w, you will also want to increase the current rating of fuse F1.

Keep in mind that the triac is rated at 6 amps, although a larger one can be substituted, if necessary. The triac should have a heatsink attached unless the load you are controlling is very light.

In the schematic of figure 1, we show the value of R7 as 680K. On our prototype, we used a small metal plate as our touch sensor, and this value for R7 worked fine. If you plan to attach the input lead to the metal part of a lamp so that the dimmer will respond when you touch the lamp itself, you will have to decrease R7 considerably. For a lamp, a good starting point is about 220K. Reduce it further still if the circuit triggers itself or seems to "lock-up" when you attach the lead to the lamp body. We do not recommend that you do this, because any touch sensor circuit will be more reliable as you reduce the size of the object on its input, but in most cases, this will work if you really want to do it. Don't forget that you will want to connect the circuit ground to earth ground in this application, or in any situation where the sensor is very far from the circuit. The easiest way to do this is to simply use a three-wire electrical cord for power and attach the ground (green) wire to circuit ground.

If you are using a touch plate mounted on the box that encloses the circuit, as we did, an earth ground is not necessary. We did attach a wire to circuit ground, however, and ran it in a loop around the inside of the box, epoxying it in place. This enhances the sensitivity and guarantees that the sensor will respond evenly when triggered from various directions. This is not necessary, of course, on a system tied to earth ground.

For a touch plate, we used the metal cover from a smaller Radio Shack enclosure. After running a screw through one of the corner holes and through the top of the enclosure, we attached a terminal lug and nut. The wire going to the sensor circuit input can be soldered to this lug.

Additional intensity levels can be added or removed, as desired. For each level, you will need an optocoupler with its transistor driver circuit and a 100K potentiometer going from pin 4 of the MOC3010 to the junction of R2, C2, and the diac. The positive end of D5 should always connect to the first unused output, in order to generate a reset and turn off the lamp when the stepper is advanced past the last used step.

PARTS LIST

Semiconductors:

B1- 100 PIV- 1.4 amp bridge rectifier (Radio Shack #276-1152)
D1-D5- 1N4148 diode (Radio Shack #276-1122)
D6, D7- 1N4004 silicon rectifier (Radio Shack #276-1103)
Diac- ECG 6408 or equivalent-32 v break-over
Q1-Q4- 2N2222A silicon transistor (Radio Shack #276-2009)
Q5- 400 v, 6 amp triac (Radio Shack #276-1000)
OC1-OC3- MOC3010 optocoupler (Radio Shack #276-134)
U1- 4001B quad NOR gate (Radio Shack #276-2401)
U2- 4013B dual D flip-flop (Radio Shack #276-2413)
U3- 4017B decade counter (Radio Shack #276-2417)
U4- 7812 12 v regulator IC (Radio Shack #276-1771)

Capacitors:

C1- .05 ufd-400 v (or two .1 ufd-200 v units in series)
C2- .1 ufd-200 v
C3- .001 ufd-50 v disc
C4- .0047 ufd-50 v disc
C5, C6- 10 ufd-16 v electrolytic
C7- 1000 ufd-25 v electrolytic
C8, C10- .1 ufd-50 v disc
C9- 100 ufd-16 v electrolytic

Resistors: (All resistors 1/4 w unless stated otherwise)

R1- 180 ohm-1/2 w
R2- 100K-1/2 w
R3- 100K trimmer potentiometer
R4- 1K-1/2 w
R5, R8- 1 meg
R6- 4.7K
R7- 680K (or a 470K and 220K resistor in series)
R9, R11- 22K
R10- 47 ohm
R12, R18- 2.2K
R13- 47K
R14-R17- 15K

Miscellaneous:

F1- 2 amp fuse and holder
F2- 1/2 amp fuse and holder

L1- 100 uH RF choke (Radio Shack #273-102)
T1- 12.6 v-450 mA transformer (Radio Shack #273-1365)
V1- varistor (Radio Shack #276-570)
Circuit board for DC components (we used a Radio Shack #276-168A)
Circuit board for AC components (we used 1/2 of a Radio Shack #276-159)
Plastic enclosure (we used a Radio Shack #270-224)
Heat sink for triac (Radio Shack #276-1363)
Circuit board standoffs, rubber feet for case, line cord,misc.hardware

BRINGING UP THE UNIT

If the touch sensor for this project is going to be a lamp or other large metal object, leave it disconnected for the moment, with a wire a foot long or so attached to the touch sensor input for testing. A jumper lead will do fine. Plug the lamp to be controlled into the dimmer circuit, making sure the lamp's switch is turned on.

Before powering up the unit, take a continuity check between both prongs of the line cord and DC circuit ground. You should read infinite resistance. If not, find out where the problem is before proceeding. Once this is resolved, plug the unit in. The lamp should remain off at this point if everything is working properly.

After power-up, cautiously touch the input to the touch sensor. Obviously, you should not feel a shock or strong tingle. In the unlikely event that you do, unplug the unit and locate the cause.

The state of the lamp after this first touch is unpredictable, since we have not yet adjusted R3. The lamp may or may not come on. Touch the input again, however, and the light should come on to a medium brightness. One more touch should turn it on to full intensity. A fourth touch should turn it off.

If the circuit has responded as it should have, touch it once again to select the first level of intensity. Keeping in mind that the potentiometer connections are at line potential, carefully adjust R3 until the lamp is at the desired brightness. Once everything is operating properly, you can then restore the connection to the touch sensor input, if it was previously disconnected. If the circuit no longer responds to your touch when you do this, you will need to reduce the value of R7 until it operates properly.

If the unit does not seem to function properly, check the following:

- Measure the power supply voltages. There should be at least 16 v across C7. If the voltage is very low or zero, check the connections on transformer T1 and bridge rectifier B1, and make sure that fuse F2 has not blown. If considerable voltage is there but it's still lower than 15-16 v, make certain that your power supply is not shorted on the 12 v side of the regulator. This regulator has built-in short circuit protection, which can limit the current flow without blowing F2. If there is at least 16 v across C7, check the voltage across C9. There should be very close to 12 v here. If the voltage across C7 is correct, but the voltage across C9 is incorrect, once again verify that there is not a short across the 12 v supply line. If there isn't, regulator U4 should be replaced.

- If the power supply voltages are correct, but the unit still does not function, short the collector of Q4 to DC circuit ground. The lamp should come on at full intensity. If it does not, you more than likely have a problem in the AC driver circuit. To make certain, remove the short from Q4's collector and measure the voltage from the collector to ground. There should be at least 14 v there. If not, check the wiring of R4 and the optocouplers. If 14 v or more is present on the collector of Q4, but shorting the collector to ground does not turn on the light, carefully short pins 4 and 6 of MOC3010 #1, remembering that these connections are running at 120 v. If the light now comes on, the MOC3010 is either bad or wired incorrectly. If the light still does not come on, measure the voltage across triac Q5. There should be 120 v here. If there is not, make sure fuse F1 has not blown, and that the lamp you are controlling has its switch turned on. Also check the wiring of L1. If 120 v is across the triac, check the wiring of R1 through optocoupler #1 to the triac's gate. If everything appears correct, replace the triac.
- Once operation of the AC driver circuit has been verified for full intensity, operation of the 50% and dim levels can be checked. Grounding the collector of Q3 should cause the lamp to turn on at about medium brightness. If the lamp worked at full intensity but not at this level, either MOC3010 #2 or D6 is bad or has been wired incorrectly. Grounding the collector of Q2 activates the dim level. If the other two levels work but this one does not, first try rotating R3 to both extremes. If the light still will not come on, short out R2 and try it again. If it works now, leave R2 shorted out or remove it from the board. You may also have to do this if the lamp will not adjust as bright as you would like for this level. Normally, this should not happen, but if you substituted a different diac, the firing voltage may be different. If it is higher, R2 may have to be reduced or removed to give R3 the correct adjustment range. At the other extreme, if you happen upon a diac with a very low trigger voltage, you may need to replace R2 with a 150K or 180K resistor in order to get R3 to adjust the lamp intensity low enough. If you have tried this and the third level still will not work, double check your wiring. If a problem can not be found, either MOC3010 #3 or the diac is bad. A voltage measurement on pin 4 of the optocoupler with Q2's collector shorted to DC ground should tell you which one is at fault. One other possibility is the potentiometer. Occasionally, you will get one that is open. With the power removed, a check with an ohmmeter will quickly verify this.

- Once all three channels of the AC driver circuit are functioning, we can troubleshoot the DC control circuit. Start by checking the supply voltages across the ICs to make sure there isn't a wiring problem between the power supply and the control circuit. On the 4001 and 4013, you should measure 12 v across pins 14 (+) and 7 (-). On the 4017, you should measure 12 v across pins 16 (+) and 8 (-). If these voltages are correct, measure the voltage on pin 11 of the 4013 IC. This is the input to the one-shot from the touch sensor. This measurement should be very close to ground, except when the input to the capacitive switch is touched, at which time it should go high (close to 12 v). If it does not go high when you touch the input, or stays high even when your hand is removed from the sensor, the problem is in the touch sensor circuit. You can refer to the chapter on the 'Touch- operated Switch' for specifics on how to troubleshoot this portion of the circuit. If the output of the touch circuit is always high, once again verify that there is not too much metal attached to the sensor input for the selected value of R7.
- If the touch sensor circuit responds correctly, measure the voltage at pin 14 of the 4017 stepper IC. This pin is the input to the 4017 from the one-shot stage. This point should normally be close to 0 v, going high whenever the sensor is touched. It should stay high as long as the sensor is being activated. When the touch sensor is released, it should stay high momentarily (about 1/5 second), and then go low. If it doesn't do this, and the input to the one-shot measured correctly, then the problem is in the one-shot stage. Check all of your wiring carefully around U2, the 4013.
- If the one-shot is correct, but the circuit still will not function, the problem is in the stepper circuit. Check all wiring around the 4017 carefully. If the stepper is getting pulses on pin 14 from the one-shot, but output 0 (pin 3) is always high, measure the voltage on pin 15 (the reset terminal). There should be less than a volt here. If there are several volts on this pin, replace C6, and check your wiring at this point. If all wiring and voltages in the stepper circuit seem correct but it will not increment, replace the IC. If the IC increments correctly, but one intensity level will not work unless you ground the collector of its respective transistor, either that transistor is bad or you have it wired incorrectly.

DC-CONTROLLED LIGHT DIMMER

This project involves the construction of a special form of light dimmer. Like most dimmers, it works by varying the phase angle at which the load is turned on. Unlike most other dimmers, however, the phase angle of turn on in this unit is set by inputting a DC control voltage. This voltage can vary between 0-12 v, with the lamp starting to turn on at about 3-4 v and achieving full brightness at about 9 v. The control circuit is totally isolated from the AC line, making it very safe for experimentation and very simple to interface other circuits, since no line isolation is required. Since its method of operation does not involve any large "break-over" voltages in the trigger circuit, as with diacs and neon bulbs, full range operation can be achieved with virtually no hysteresis; therefore, precise control can be maintained even at very low lighting levels.

There are many useful applications for this device, some of which are not immediately obvious. The original application for which this circuit was designed was a night light for a child who did not want the light turned off at bedtime. A simple timer could have been used to turn the light off after he fell asleep, but it was desired to have the lamp stay on as a night light, only at a much lower level. Rather than have the light dim in a sudden transition, we used this circuit with a capacitor across the input. Once charged to +12 v, it slowly discharged through a resistor to gradually dim the light to a preset level over a period of about 30 minutes. This "fading" effect can be used for other purposes by changing the values of the timing capacitor and resistor. Relatively small values can cause a faster fade of 5-10 seconds, producing a theatrical effect. Values can be selected to cause a lamp to fade to a preset lower level over a 5-10 minute period, perfect for the bachelor who wants to dim the lights on his date without being obvious.

The reverse effect can also be created by charging the capacitor through the resistor. Special lighting effects (especially using flood lamps) can be created by inputting different waveforms, such as low frequency (1 Hz or less) sine waves. Since our input is a DC voltage, it is also relatively easy to provide feedback to regulate the light output. We can use a photocell to provide feedback to the input so that it will turn the light on in a room in increasing increments as it gets darker outside, compensating by the correct amount to always keep the light constant.

By using a thermistor as our sensing element and a heat lamp as our load, we can also use this circuit as a precision temperature regulator.

We can also connect a touch-operated switch, with proper interfacing circuitry, to this project to create a touch-activated dimmer. And, of course, it also makes an excellent standard light dimmer, with very precise control. We will show you how this circuit can be used to accomplish all of these things.

CIRCUIT DESCRIPTION

Looking at figure 1, we see that this circuit can be broken down into three basic parts: a power supply, a DC control circuit, and an AC driver circuit. The power supply is a typical 12 v DC supply. The AC voltage from the power line is reduced to about 12.6 v by transformer T1. Bridge B1 converts this to a pulsating DC waveform. After passing through D1, capacitor C2 filters this to give us a relatively smooth DC voltage. We need D1 here because our control circuit needs the 120 Hz pulsating DC voltage leaving the bridge rectifier for its timing. If D1 were not there, all the control circuit would see is a steady DC voltage. Resistor R1 and zener diode D2 regulate our power supply voltage to about 12 v for the control circuit, and capacitors C3 and C4 provide additional filtering.

If you want to use this project in an application where small size is of importance, you might be tempted to remove the transformer and use a couple of resistors to drop the line voltage. This will work, but you will lose your AC line isolation, causing a potential shock hazard. Naturally, you would not use this technique if you are building a touch-operated dimmer! You must also watch the power ratings on the resistors. In the interest of safety, we do not recommend this practice.

The DC control circuit is centered around a 1458 dual op-amp. This IC includes two op-amps similar to the 741 in one package. The first op-amp ("A") has its inverting (-) input connected to the 120 Hz pulsating DC voltage leaving the bridge rectifier through the network composed of R2, R3, and D3. This network serves to protect the input, as the peak voltage being applied (about 18 v) will exceed the power supply voltage of the op-amp. D3 prevents the voltage at the input from ever rising above 6 v.

This gives us a clipped 6 v signal dropping to 0 v one hundred twenty times per second on the inverting input. This drop to 0 v is in sync with the power line voltage as it crosses the zero-axis and provides our circuit with a synchronization signal. For proper operation, the 120 v driving the transformer must be from the same source as that driving the triac circuit, or the trigger pulses to the triac will not arrive at the proper time.

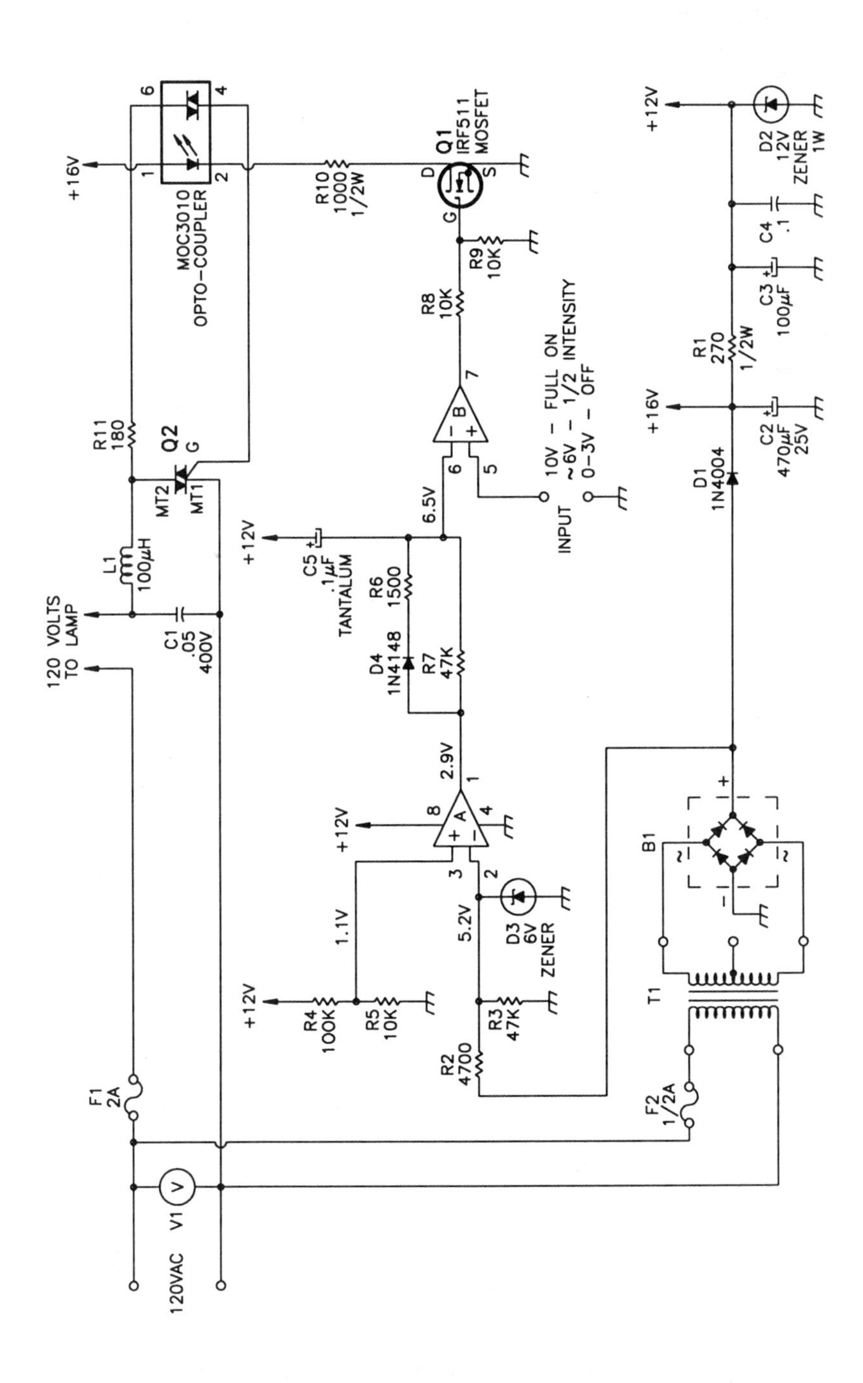

Figure 1 Schematic of the 'DC-controlled Light Dimmer.'

The non-inverting (+) input of op-amp "A" is connected to R4 and R5, forming a voltage divider and setting the potential at this input to about 1.1 v. Feedback is not connected from the output to either input of this op-amp; it is running at full gain as a comparator; therefore, the output will almost always be either entirely on or off, depending upon the input voltages at that moment. Since the voltage on the non-invert

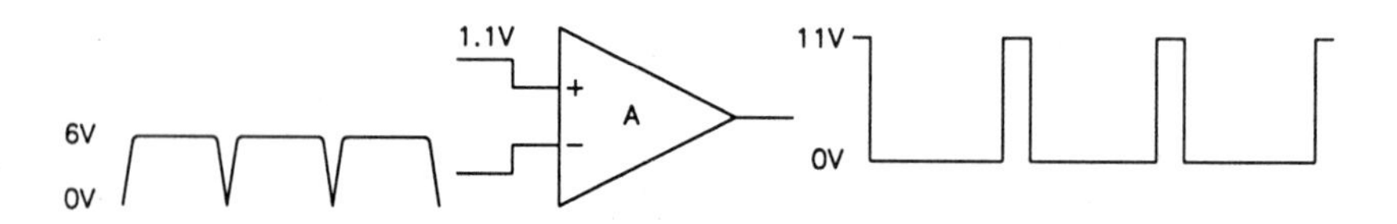

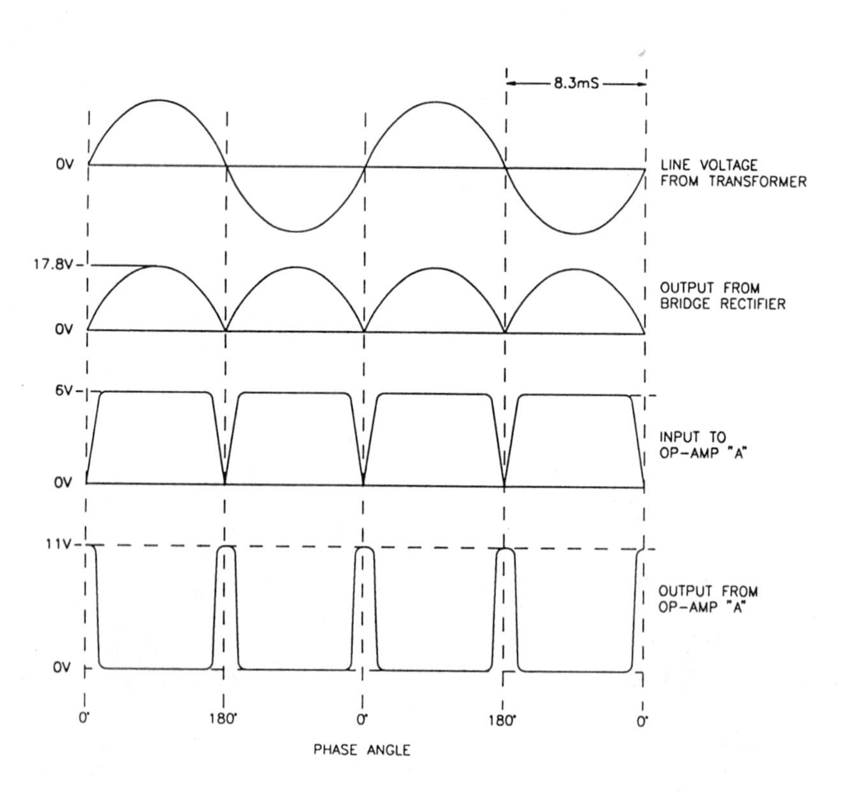

Figure 2 Input and output relationship of op-amp "A."

ing input is fixed by R4 and R5, the output will depend upon whether the voltage on the inverting input is above or below 1.1 v, and since this input is following the pulsating voltage from the bridge, the output of the op-amp will be low most of the time, going high momentarily only when the power line voltage is crossing zero. Figure 2 shows the relationship between the input and output of this circuit.

On the output of this circuit, we have capacitor C5 being charged through resistor R7 when the output swings low. With the values for C5 and R7 as shown, the voltage on the inverting input of the second op-amp will gradually drop from about 11 v to around 3.5-4 v during each cycle. As the AC line approaches the zero- axis, the output of op-amp "A" goes positive. In order to discharge C5 quickly to prepare for the next cycle, D4 and R6 have been added across R7.

The purpose of this has been to provide a relatively linear waveform, where the signal "sweeps down" several volts for each half-cycle of the AC line voltage (see figure 3). Using the first amplifier to square up the signal from the bridge rectifier gives us a more linear sweep than if we had connected the RC network directly to a sine waveform. The sweep is still not linear, but it is close enough for what we are doing.

The output of the second op-amp ("B") is connected to a MOSFET, Q1, which in turn drives an optotriac. When the output of op-amp "B" swings positive, Q1 turns on the optotriac. This triggers the power triac, Q2, turning it on; therefore, the power triac will be turned on whenever the non-inverting (+) input of op-amp "B" is more positive than the inverting (-) input.

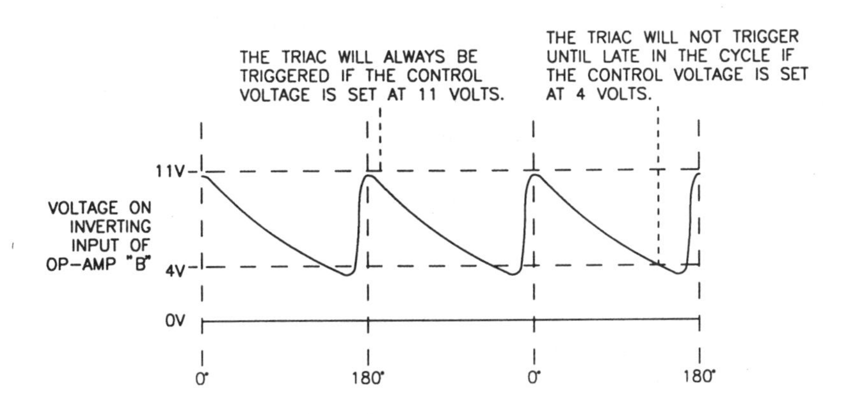

Figure 3 Waveform on the inverting input of op-amp "B."

We see from figure 3 that if the non-inverting input is held below 3 v, the output will never go positive because the waveform on the inverting input never sweeps that low. As a result, the triac is never triggered, and the lamp connected to the circuit remains off. If the non-inverting input is held at a level of 11 v or more, the triac is always on, since this input will always be more positive than the inverting input. As a result, the lamp that we are controlling will run at full brightness. Notice, though, that if we gradually lower this voltage, we reach a point where the inverting input has a higher potential on it than the non-inverting input early in the cycle. This delays the cutting on of the triac, dimming the controlled lamp slightly. As we lower the control voltage further, the inverting input remains higher in potential for a greater part of the cycle, retarding the triac turn on still more. As the control voltage reaches about 4 v, the triac is being turned on so late in the cycle that the lamp will barely light at all. We, therefore, have complete control of our lamp intensity by simply putting our control voltage where we want it.

The AC driver circuit, consisting of the triac, optotriac, R11, L1, and C1, is a simple solid-state switch. When the LED in the optocoupler is turned on by Q1, the photosensitive triac internal to the MOC3010 turns on, triggering triac Q2 through R11. This triac will then remain on until the power line voltage crosses the zero-axis. At this point, the current flow through the triac will be zero, and it will shut off, provided the optotriac is also off. The purpose of L1 and C1 is to help suppress any RF interference created by the triac switching on in mid-cycle. The varistor has been added to protect the circuit from voltage spikes on the power line. It is not essential for proper operation, but it is good insurance, particularly if the device will be used in an application where it will remain on all the time, such as a touch dimmer.

We have fused this circuit at 2 amps because of L1 (the part we used here is only rated to carry that much). This would correspond to a rating of 240 w. If you wish to drive a heavier load, a heavier coil can be substituted; the triac will handle up to 6 amp. A heavier triac can also be substituted, if desired. If a choke coil with a heavier current rating is not readily available, two can be used in parallel. This will nearly double your current capacity. It will also reduce the inductance, but it should still be satisfactory. As a last resort, L1 can be totally removed from the circuit and replaced with a wire of sufficient size, provided the RF interference is not a problem. C1 will help to reduce the RF interference a lot by itself. Do not forget to also increase the rating of the F1 fuse. It must be large enough to handle the current that your load will be drawing.

To make this circuit function as a standard dimmer, all we have to do is add a potentiometer (see figure 4). This will vary the input voltage from 0-12 v as the control is rotated. The value of the potentiometer is not critical; anything from 10K to 1 meg would be suitable. Just make sure it has a linear taper and not an audio taper (an audio taper potentiometer is designed for volume controls and has a response to match that of the human ear. The resistance will be compressed at one end of the rotation, making it difficult to use here).

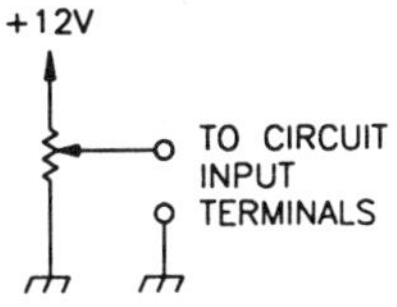

Figure 4 Input for use as a standard light dimmer.

This dimmer will usually give better performance at both the high and low end of its range than the traditional diac or neon bulb-type dimmer. The reason is that the load can be turned on over a wider range of phase angles. Figure 5 shows a simple light dimmer circuit using a neon bulb. As the AC waveform goes high, capacitor C starts to charge through resistor R. When, and if, the charge on C reaches the breakdown voltage of the neon lamp, C partially discharges through it to trigger the triac. By varying R, the rate at which C charges will change, and you will vary the phase angle at which triggering occurs. Notice that since the breakdown voltage of the neon lamp is often 80-100 v, triggering will never occur before the voltage has reached this level, regardless of how small we make R. This limits the maximum brightness that can be achieved. We also have the same problem at the other end of the half-cycle. If C has not reached the break-over point before the waveform falls to 90 v, it will never get there; therefore, we also have mediocre performance at the dim end of the range.

This problem can be overcome to a large degree by replacing the neon bulb with a diac. This is a solid-state component which has a breakdown characteristic quite similar to that of a neon bulb, except that the break-over voltage is generally in the 30-40 v range. This lower break-over voltage broadens the range of phase angle turn on.

This DC-controlled dimmer can do even better, though it is not totally immune from this problem. Average turn on will be in the 22-24 v range because 1.1 v will not appear on the inverting input of op-amp "A" until the transformer secondary is putting out about 2.4 v (there is a small voltage drop across R2 and across the two conducting diodes in the bridge). By comparing this instantaneous voltage to our peak voltage (17.8 v) and using our sine tables, we find that op-amp "A" will switch low at about 8° and high at about 172°. This means that the output is high for a period of about 370 microseconds both before and after the zero-crossing.

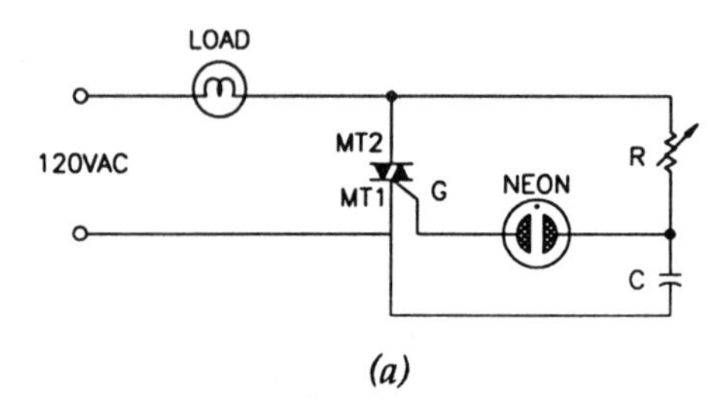

(a)

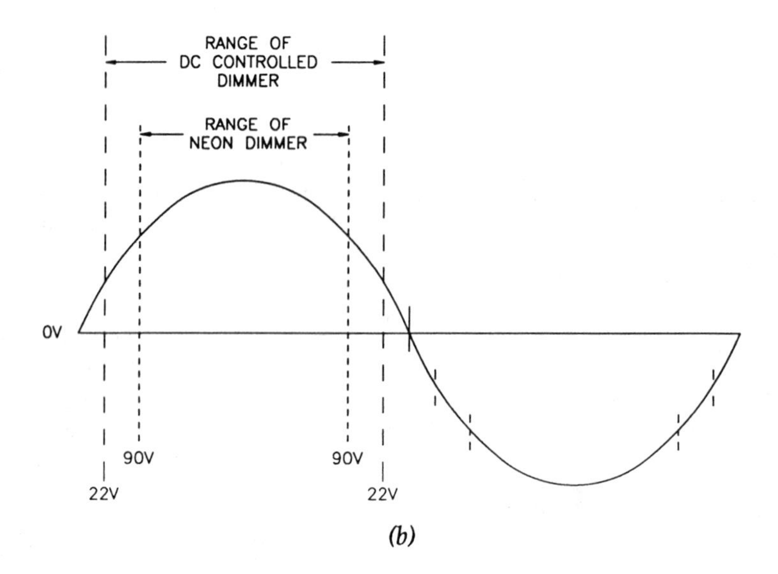

(b)

Figure 5 (a) A simple light dimmer using a neon bulb as the threshold device; (b) a comparison of the turn-on phase angle using the neon dimmer and the DC-controlled dimmer circuit.

We could reduce this even more by lowering the voltage on the non-inverting input of op-amp "A", but there is a reason why we do not want to do this. In the neon-triggered dimmer, the timing capacitor C was discharged each cycle as it fired the triac. Even if the break-over voltage was not reached, it would discharge through resistor R on the next half-cycle, since the polarity would then be reversed. In our circuit, C5 continues to charge, regardless of when firing occurs; therefore, we must force the output of op-amp "A" high early enough in the cycle to allow C5 time to discharge through D4 and R6 before the zero-crossing. Otherwise, the optotriac will still be turned on at this time and trigger the triac at the start of the next cycle as well, turning the light on at full intensity. The trigger point that we have selected allows adequate time for D4 and R6 to discharge C5, while still giving maximum control range. This dimmer will take the lighting down to the point where the filament is just starting to glow. We see no reason to go beyond this.

Figure 6 shows interface circuits for some applications of this project. Figure 6(a) shows a circuit with a simple RC combination connected to the input. Closing the switch charges the capacitor, turning the light on at full intensity. When the switch is opened, the capacitor will slowly discharge through variable resistor R, causing the lamp to slowly fade until it is completely off. The time delay until the lamp goes off is very close to the value of C in microfarads multiplied by R in megohms. Figure 7 shows some of the time delays we got with our prototype.

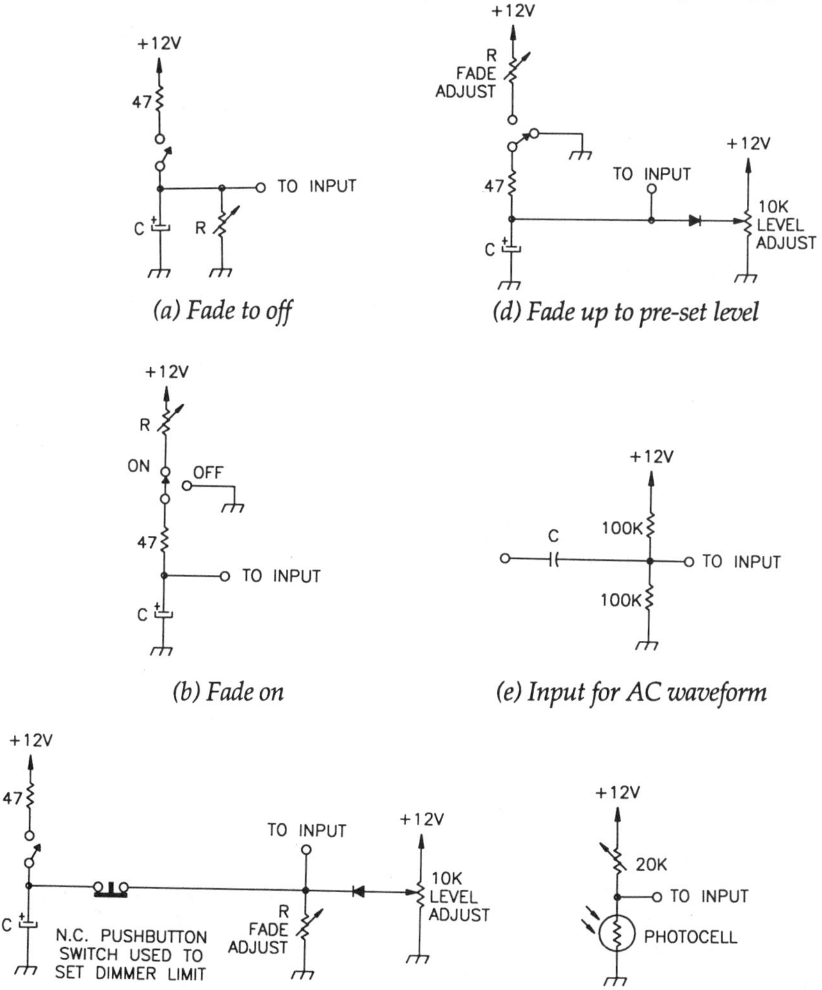

(a) Fade to off

(d) Fade up to pre-set level

(b) Fade on

(e) Input for AC waveform

(c) Fade down to pre-set level

(f) Input for light control

Figure 6 Various interfaces that can be used with the DC-controlled dimmer circuit.

Figure 6(b) shows an input to get the reverse effect. Moving the switch to the "ON" position will gradually raise the light's intensity. Moving the swich to the "OFF" position will turn it off instantly. Placing a resistor on the "OFF" terminal of the switch to ground will allow the lamp to fade both on and off.

Figure 6(c) shows a circuit that does the same thing as 6(a), except that a level control has been added, which will stop the fading at a certain point. Figure 6(d) shows a circuit that does the same thing for 6(b). In both cases, best results are obtained if the resistance of the potentiometer is considerably lower than that of the timing resistor. A 10K potentiometer is often a good choice.

Figure 6(e) shows a possible circuit for inputting a sine wave to create an effect of slowly brightening and dimming the lamp. Since the input waveform would be at a very low frequency (usually 1 Hz or less), the capacitor should be rather large. Other waveforms can be input for other effects, often without this interface. A 9-12 v squarewave can be directly inputted, turning the lamp into a flasher. Figure 6(f) shows a photocell connected to the input to gradually turn on the light as it gets dark outside.

Capacitor	Resistor	Time*
100 ufd.	47K	5 sec.
100 ufd.	100K	10 sec.
100 ufd.	10M	18 min.
470 ufd.	100K	1 min.
470 ufd.	470K	5 min.

*Full-on to Full-off

Figure 7 Typical time periods for various RC combinations with a 75 w load.

Another useful application of this circuit is a precision temperature controller. We use a heat lamp, or possibly even a regular light bulb, as our heat source. Rather than cutting the lamp on and off with a thermostat, we run the lamp all the time, but dim it to the level where the heat it generates balances the heat loss of our system. Naturally, this would only be used for applications requiring a moderate amount of heat and very precise regulation. This circuit could be used to build a small experimental heat chamber out of a large styrofoam cooler, or possibly used to help keep some baby chickens warm. You would not want to use it for running an electric space heater. Although a larger triac could handle the current demand through the heating unit, the type of fan motor used in these devices would likely be destroyed by the method we are using.

We might be tempted at first to try building one of these by replacing the photocell of figure 6(f) with a thermistor. This would cause the voltage on our circuit input to rise as the temperature decreased, turning the lamp on brighter. Actually, this will work to an extent, but there is a problem. With the circuit in figure 6(f), the thermistor is producing a change on our circuit input of only 10 mV per degree of change, while the control voltage must cover a range of about 6 v to take the lamp from full on to full off; therefore, if you were to try this circuit, it would probably not remain very stable. As the temperature started to cool, the thermistor would try to turn the lamp on more fully, but the temperature would have to drop a couple of degrees or more before the lamp would be turned on enough to compensate. Then the process could reverse and go to the other extreme. You don't end up with a precision regulator; what you do end up with is a circuit that has two limits that it will not exceed (possibly several degrees apart), with the temperature wandering around somewhere in between. This is not what we wanted.

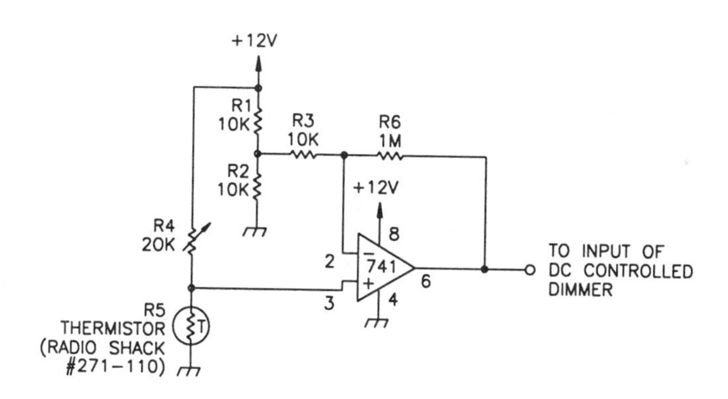

Figure 8 An input circuit to use the DC-controlled dimmer as a precision temperature controller.

The solution to our problem is to add some gain between our sensor and the input to the DC-controlled dimmer. This will cause a small variation in temperature to produce a substantial change in the input voltage. Figure 8 shows a suitable circuit using a 741 op-amp. The inverting input is held at 1/2 the supply voltage by R1 and R2. They are connected to the input by R3, and R6 serves to provide negative feedback. The gain of the circuit is determined by R3 and R6, and it will not be the exact ratio of R6/R3, since we are using the op-amp in the non-inverting mode, but this is a close approximation, as long as R6 is considerably larger than R3. On the non-inverting input, we have a thermistor and a potentiometer forming a voltage divider, just as we discussed before. But now, any change of voltage in the thermistor/potentiometer divider will be multiplied by approximately 100. If a change of 1/4°F might have produced a change of 20 mV on the input before, it will now produce a change of about 2 v. This results in a circuit that is useful for any situation requiring very tight temperature regulation.

This same circuit is also useful for regulating the amount of light that falls on a certain object. If you need to keep the light intensity constant, a photocell can be substituted for the thermistor. In the circuit of figure 6(f), we used a photocell to increase the lamp intensity as the room got darker. While this approach would probably be fine for most lighting applications, the change may not be enough to guarantee true regulation, although the change in resistance of a photocell is usually somewhat more dramatic than that of a thermistor. In situations where the light intensity must be held very tightly, as in some science experiments, this circuit can be used. You will probably want to reduce the gain of the circuit (by reducing R6), since the response of a photocell is generally superior to that of a thermistor. Too much gain may make the circuit "jumpy," causing the lamp to oscillate between full on and full off, due to the fact that the dimmer will often respond much faster to changes in the photocell's resistance than the photocell can respond to changes in the light output; thus, the circuit may tend to slightly overcompensate. If there is too much gain in the system, this slight overcompensation will no longer be slight. A moderate amount of gain will give close regulation without sacrificing stability.

If you decide to use this circuit as a precision temperature regulator, there are a couple of things to keep in mind. One is that the useful range of the temperature control potentiometer, R4, will narrow as you increase the gain of this circuit. Because of the increased gain, the controlled lamp will be totally on or off for most of the potentiometer's range, dimming only over a very narrow part of the control's rotation. To obtain more resolution for a precise temperature setting, you might want to find the value of your thermistor at the temperature you wish regulated, and then replace the potentiometer with the largest standard resistor size less than this value, in series with a potentiometer of relatively low resistance. As an example, suppose your thermistor has a value of 5.2K at the temperature you are interested in. If you remove the 20K potentiometer and replace it with a 4.7K resistor in series with a 1K potentiometer, you will have narrowed down the adjustable temperature range, but greatly increased the resolution within that range, allowing for very precise settings. A gain of one hundred for this circuit is not magical. You can increase the value of R6 for still more gain if you wish, or decrease it for less gain. If you increase the gain too much, the lamp will spend most of its time either fully on or fully off, going from one extreme to the other, trying desperately to keep the temperature regulated to within 1/1000th of a degree. As with the photocell, a moderate amount of gain will generally give the best performance.

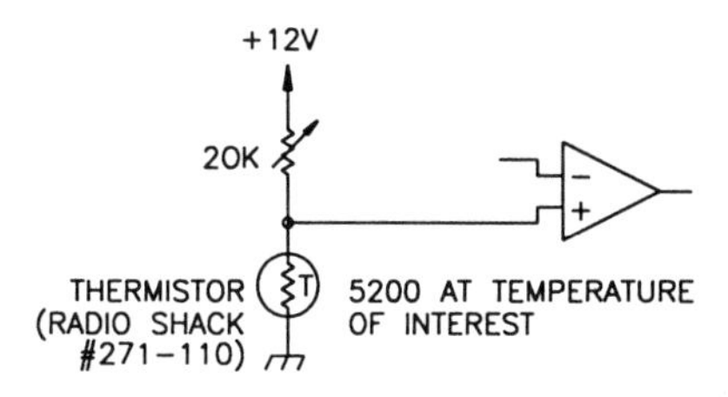

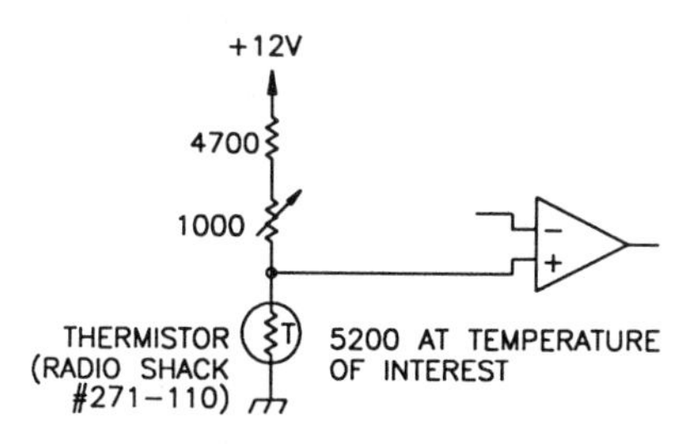

Figure 9 Increasing the resolution of the temperature control.

The other detail to watch here is the rating of the lamp that you use. Regular light bulbs may provide sufficient heat for some uses, but the 250 w heat lamps are the most obvious choice for this application. Since the output circuit, as shown in the schematic, is only rated to carry 2 amp (240 w), as mentioned previously, make sure to use either a large inductor, two inductors, or to remove L1 altogether. The fuse will also have to be increased to 3 amp.

One other useful application for the DC-controlled dimmer is to make your own touch-controlled dimmer (see figure 10). We start by taking a touch-operated switch (see the 'Touch-operated Switch' project for details on this) and interfacing it to a 4017 decade counter through a monostable (one-shot) circuit. The 4013 one-shot is slightly modified from that shown in the 'Touch-operated Switch' project. In that application, the 4013 output is a 2/3-1 second pulse every time the sensor plate is touched. This is fine for on and off control, but for a touch dimmer, where we might want to touch the plate 2-3 times in rapid succession, we need a "semi-automatic" capability. With the circuit as shown, touching the sensor triggers the flip-flop, but it also turns on the 2N2222 transistor, which shorts out the timing capacitor; therefore, the circuit will not start to time out until the plate is released. Once it is released, the circuit will time out in less than 1/5 second. This allows a faster recovery time for multiple touches, while still guaranteeing one, and only one, pulse out per touch.

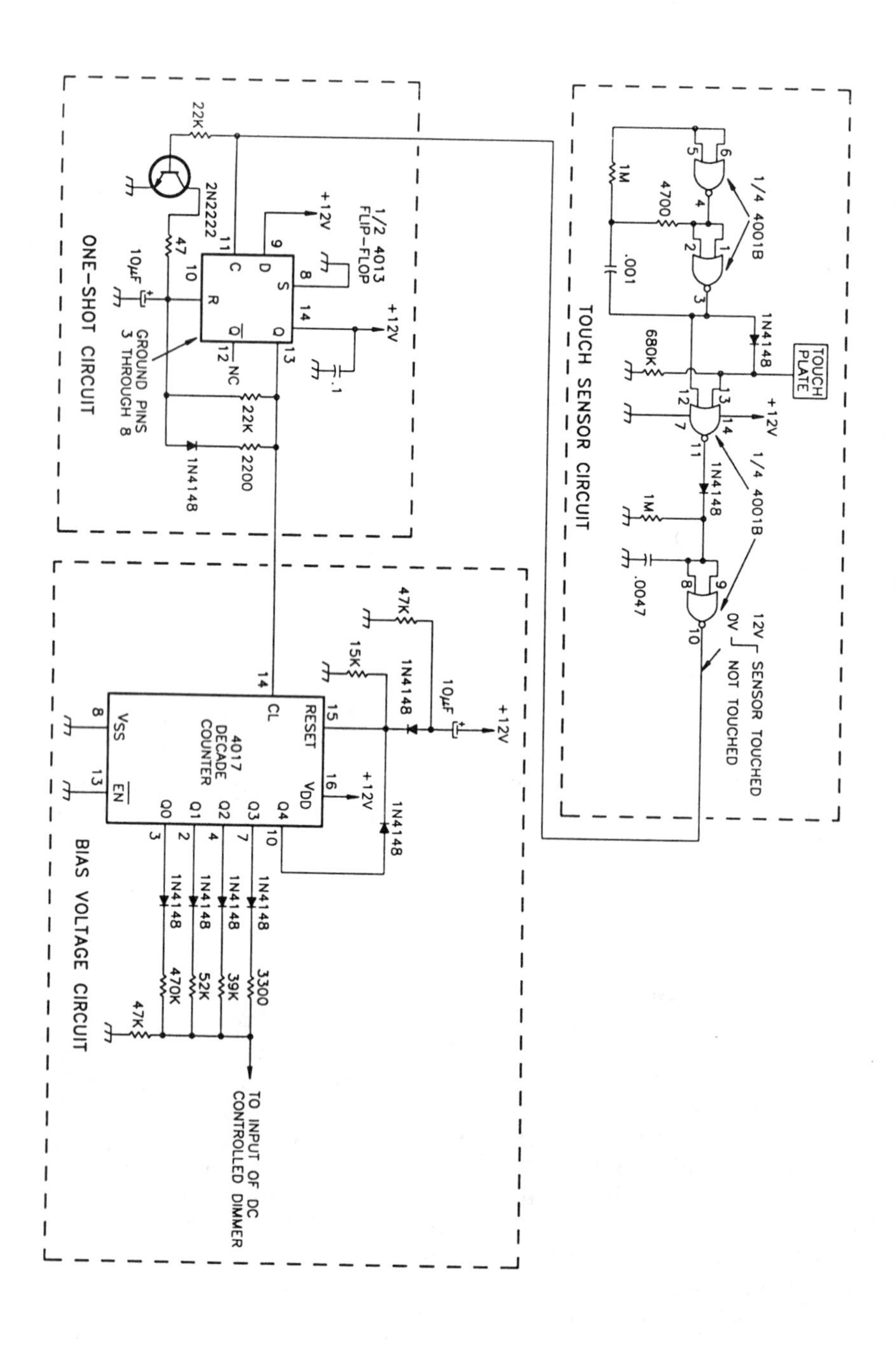

Figure 10 Interface for a touch dimmer circuit.

The 4017 IC is a decade counter configured with ten separate outputs, only one of which is high at any time. Each touch of the sensor plate will cause the one-shot to output one clock pulse to the 4017, incrementing the count to the next output. When a particular output is high, the resistor on that output forms a voltage divider with the 47K resistor (see figure 10), changing the control voltage inputted to the dimmer circuit. The diodes shown in line with the resistors on each output effectively remove all of the resistors on the non-selected outputs from the divider. On any low pin, the diode will be reverse biased (for more details on the operation of the 4017, see the chapter on the 'Light Sequencer').

In the circuit of figure 10, when output 0 is high, we have a diode in series with a 470K resistor and a 47K resistor. The input to the DC controller will have a potential of about 1 v on it, keeping the lamp off. The resistors on the other outputs will have no effect. When the next clock pulse arrives, output 0 goes low and output 1 goes high. We now have a 52K resistor in series with a diode and the 47K resistor, giving us a control voltage of about 4.5-5 v. The next clock pulse will raise the control voltage to about 6 v, the third pulse will raise the voltage to over 10 v, giving us full intensity, and the fourth pulse will raise output 4 high. Note that we have a diode connected from this output to the reset terminal. This will reset the device and output 0 will again go high, turning off the lamp.

Since the 4017 has ten outputs, up to nine steps in the lighting can be achieved (the remaining output is used for "off"). Simply add a new diode and resistor for each new step, and move the positive end of the diode currently on output 4 to the first unused output (the diode is eliminated if all ten outputs are used).

The values of the resistors shown on these outputs are somewhat arbitrary. The 3.3K resistor, used for full intensity, and the 470K resistor, used for full off, would not generally be varied, but for all of the intermediate steps, feel free to change the values to whatever you want. Probably the best idea is to use a 100K trimmer potentiometer in place of the fixed resistor for the middle levels, making it simple to adjust the light intensity as each step is selected. In our prototype, a resistance of 52K for the first step was a good choice for us, giving a very dim light. A 47K resistor may not be too bright. For the next step, a 39K resistor gave a good moderate light output. There is bound to be some slight variation from unit to unit, as well as in personal preference, so we strongly recommend that you use a variable resistor for these middle levels.

CONSTRUCTION DETAILS

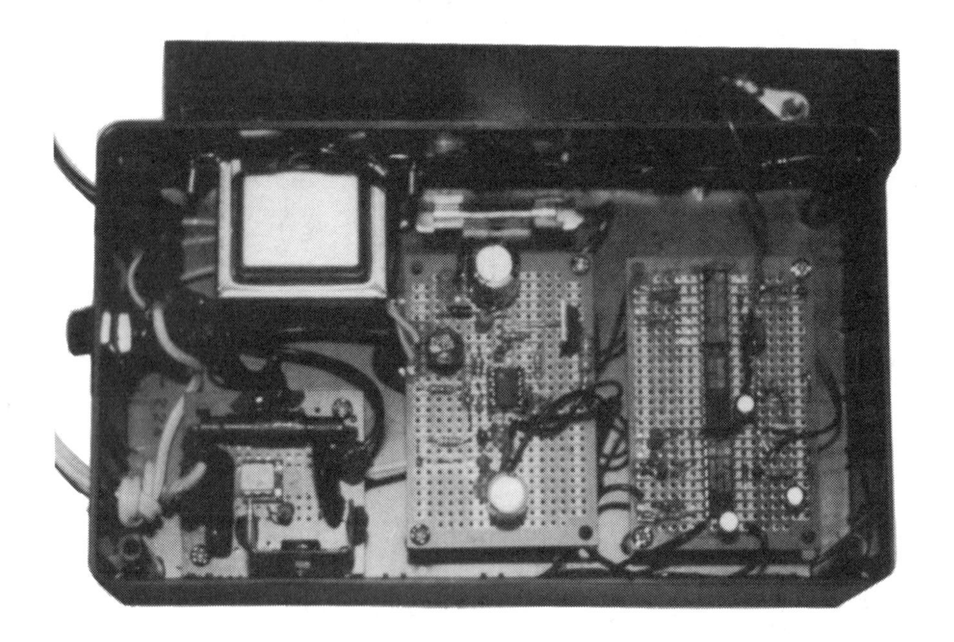

Interior view of a completed DC-controlled light dimmer project.

Although all of the components for this project could be mounted on one board, we decided to take a modular approach. We assembled all of the line voltage components (the triac, varistor, L1, C1, and the optocoupler) on one-half of a Radio Shack board (#276-159). We felt that this would help ensure that the AC line voltage would not accidently get into the DC circuitry. We then mounted the DC control circuitry and power supply components on another Radio Shack board (#276-150). Depending upon your application, and the additional circuitry you may need, you may wish to use a larger board. We decided to stick with the modular route, and use an additional board for any input circuitry. This allowed us to easily use the same circuit for various tests and applications. Mounted in a Radio Shack case (#270-224), we had plenty of room for all the boards. You may wish to fit any additional interface circuitry on the DC control board and use a smaller case, if you only have one application in mind.

If any additional components added to the input will be drawing a substantial amount of current from the regulated 12 v supply, R1 may have to be reduced in value. Another alternative is to eliminate R1 and D2 and replace them with a 12 v regulator IC. The power supply as shown should be able to easily handle an additional 10 mA with no problems. With CMOS, this can go a long way. The entire touch dimmer interface shown in figure 10 can easily run off this.

A plastic case is probably best for this project, since some of the circuitry involves 120 v. Make absolutely certain that the AC is not getting into the DC control circuitry in any way. When you are finished building this, a continuity check between either terminal of the line plug and ground of the DC circuitry should show infinite resistance. Although you can buy line cords for projects like this, when you have 120 v leaving the device to power a load, it is often far simpler and cheaper to buy an extension cord with molded ends and cut it to the proper lengths. You can also mount a small 120 v receptable on the project case itself.

PARTS LIST

Semiconductors:

B1- 100PIV-1.4 amp bridge rectifier (Radio Shack #276-1152)
D1- 1N4004 silicon rectifier-400 PIV-1 amp (Radio Shack #276-1103)
D2- 1N4742-12 v zener diode-1 w (Radio Shack #276-563)
D3- 1N4735-6 v zener diode-1 w (Radio Shack #276-561)
D4- 1N4148 signal diode (Radio Shack #276-1122)
IC1- 1458 dual op-amp (Radio Shack #276-038)
OC1- MOC3010 optocoupler (Radio Shack #276-134)
Q1- IRF511 MOSFET (Radio Shack #276-2072)
Q2- triac-400 PIV-6 amp (Radio Shack #276-1000)

Capacitors:

C1- .05 ufd-400 v (two .1 ufd-200v capacitors can be used in series if you have difficulty obtaining a 400 v unit)
C2- 470 ufd-25 v minimum electrolytic
C3- 100 ufd-16 v minimum electrolytic
C4- .1 ufd-50 v disc
C5- .1 ufd-16 v tantalum

Resistors: (All resistors 1/4 w unless stated otherwise)

R1- 270 ohm-1/2 w
R2- 4.7K
R3, R7- 47K
R4- 100K
R5, R8, R9- 10K
R6- 1.5K
R10- 1K-1/2 w
R11- 180 ohm-1/2 w

Miscellaneous:

F1- 2 amp fuse and holder
F2- 1/2 amp fuse and holder
L1- 100 uH RF choke (Radio Shack #273-102)
T1- 12.6 v-450 mA transformer (Radio Shack #273-1365)
V1- varistor (Radio Shack #276-570)
Circuit board for DC components (we used a Radio Shack #276-150)
Circuit board for AC components (we used one-half of a Radio Shack (#276-159)
Heat sink for triac (Radio Shack #276-1363)
Project case (we used a Radio Shack #270-224)
Power cord, circuit board standoffs, rubber feet for case, misc. hardware

BRINGING UP THE UNIT

Once the unit has been built and all the wiring double checked, make a continuity check between both prongs of the line cord and DC circuit ground. There should be no measurable leakage between the two circuits. If everything checks out, connect a potentiometer to the circuit input (as in figure 4) for initial testing. Any value from 10K-1 meg is fine. Connect a lamp to the output (with the switch turned on), and then plug the unit in. Rotating the potentiometer should cause the lamp to brighten and dim, covering the entire range from full on to full off.

If the unit does not seem to function properly, check the following:

- If the unit does not turn on the lamp at all, regardless of the potentiometer setting, start by checking the power supply voltages. There should be about 12 v across pins 4 and 8 of the 1458 dual op-amp. If not, you have a power supply problem. Next, check the voltage across C2. There should be around 16-18 v here. If this is correct, check R1 and D2. Make sure D2 is inserted in the right direction. Also make certain that there is no short across the 12 v supply rails. If there is no voltage across C2, check the wiring of the bridge, D1, and the transformer. Make sure that fuse F2 has been inserted and has not blown.
- If the power supply voltages are good, but the lamp will not come on, short the drain of the IRF511 MOSFET to ground. The light should come on at full intensity. If it does, the AC driver circuitry is working, and you can go to step 3 below. If it does not come on, make certain the optoisolator and R10 are wired properly. If they are, then you have a problem in the AC driver circuit. Keeping in mind that they are at AC line potential, carefully short pins 4 and 6 on the MOC3010 optotriac. If the light comes on when you do this, the

optotriac is probably bad. If the light still will not come on, measure the voltage across main terminals 1 and 2 of the triac, Q2. If there is 120 v here, as there should be, replace the triac. If the voltage is not here, then either fuse F1 is bad or else you have a wiring error between the triac and the power plug. Although we hate to have to mention this, if the lamp you are using as a test load has a switch, make sure it is turned on, and that the bulb is good.

- If the light does come on when you short the drain of Q1 to ground, then the AC circuitry is in working order. The next step will be to short the gate of the FET to +12 v. If the light still comes on when you do this, then the output stage is working as well, but if the light does not come on, then you either have a bad connection between the source terminal of the FET and ground, or else the IRF511 MOSFET is defective and should be replaced.
- If applying the 12 v to the gate does make the light turn on, check the input and output voltages of the op-amps, and compare your readings with those on the schematic in figure 1. Your measurements should be close to what we have. If pin 6 of the 1458 IC has around 6-7 v on it, the problem is more than likely in the second op-amp circuit (op-amp "B"). Make certain that the potentiometer you are using to change the control voltage is actually doing so. If both inputs to op-amp "B" seem correct, but the output does not switch, replace the op-amp. If the output is switching high, but the voltage is not getting to the gate of the FET, there is obviously a problem with the connections at R8 and R9; however, if the voltage on pin 6 of the 1458 seems to be way off, check the inputs to op-amp "A." There should be about 1.1 v on pin 3, and about 5-5.5 v on pin 2. If these are correct, check the network of D4, R6, R7, and C5 very carefully for errors, and make certain D4 is good. If everything seems to be correct, replace the 1458 IC.
- If the light stays on continuously, measure the voltage on the drain of Q1. If you read a voltage of 14 v or higher, the FET is off. This means that you either have a bad triac, opto-isolator, or a wiring error in the AC circuitry. Unsoldering one lead of R11 will quickly isolate whether the triac or optoisolator is at fault. If the lamp stays on after opening the connection at R11, the triac is defective. If opening R11 makes the lamp go off, the MOC3010 should be replaced.
- If the voltage on the drain of Q1 is low, then the problem is in the DC circuitry. Measure the gate voltage of the FET. If it reads 1.5 v or less while the lamp is on, replace the FET. If the FET is being turned on by the circuit (gate voltage is 3 v or higher), check the inputs to op-amp "B." Make sure that the control voltage is not staying high due to a bad potentiometer. If this seems to be correct, check the inverting input (pin 6) for 6-7 v. If this is also correct, replace the op-amp. If the pin 6 voltage is incorrect, check the inputs and output to op-amp "A" for any irregularities. Make certain that D4 is not de-

fective, nor installed backwards. If the readings on both inputs to op-amp "A" are correct, but the output stays high or low, replace the IC.

- If the light comes on, but the dimming action does not seem very linear, or is compressed toward the upper end of the range, and if the circuit seems to be working, but is responding in a non-linear way, carefully check the output of op-amp "A" and the voltage on pin 6. If the output voltage of the first op-amp is not in the ballpark, check the input voltages to see why. If the output voltage seems reasonable, but the voltage on pin 6 is off by a considerable amount, check the network of D4, R6, R7, and C5 for any problems. If nothing wrong can be found, you may have a component out of tolerance. R7 and C5 set the rate at which the voltage on pin 6 sweeps down. If the voltage on pin 6 is too high, try another capacitor for C5, or reduce the value of R7; C5 may have too much capacitance. If C5 has too little capacitance, the voltage reading on pin 6 of the 1458 will be low. The lamp will dim, but may cut off abruptly at some point. Adjust R7 and C5 so that the light just starts to come on when the control voltage is between 3-4 v.
- If the circuit shows any tendency to suddenly "jump" to full brightness as you increase the control voltage, replace D4 with a 1N4002 diode and drop R6 to 470 ohms to discharge C5 at a faster rate. If this does not solve the problem, first try replacing the optotriac and, as a last resort, decrease the value of R4 to 47K. What we are trying to do is guarantee that the MOC3010 optotriac has shut off prior to the zero-crossing of the AC line. We have never had a problem with this, however, and reducing R6 should take care of it if it does happen.
- Incidentally, varying the value of R7 may be of use in some special applications. Increasing R7 slows the charge rate of C5, and, therefore, tends to "compress" the adjustment range. The range with R7 as shown allows full control over about a 6 v spread (3.5-9.5 v). If you need to compress this spread to interface to a 5 v digital circuit or a D/A converter, increasing R7 will solve this problem; however, the sweep works in a "top-down" direction. The full intensity voltage changes very little; it is the cut-off voltage that is being moved up. You may have to offset the grounds of the two circuits, or use some other type of interface for proper operation.

FENCE CHARGER

This project is a simple fence charger circuit designed to keep livestock in, or pests out, of an enclosed area. It accomplishes this by supplying pulses to the fence at a voltage high enough to make it undesirable to touch, yet at a current level and pulse rate too low to be dangerous, assuming good sense and prudence are exercised. Naturally, we do not recommend that this device be used for any anti-social activities (i.e., attaching it to the knocker on your front door to drive away salesmen).

Our application for this project was to keep deer out of the tomatoes in a garden, a function which it performed admirably. It could also be used to protect your fruit trees and flower gardens. Although we have not used the charger for fencing livestock, it could also be used for this purpose. Keep in mind, though, that the output of this unit will probably not be as powerful as many commercial chargers, and it is intended primarily for protecting relatively small areas.

CIRCUIT DESCRIPTION

Figure 1 shows the schematic for the fence charger. This circuit is built around a simple capacitive discharge network, which pulses an auto coil at periodic intervals to produce the high voltage output. A trigger network is also included to fire the discharge network at the proper time. The circuit may be a little easier to understand by separating these two networks and looking at them one at a time.

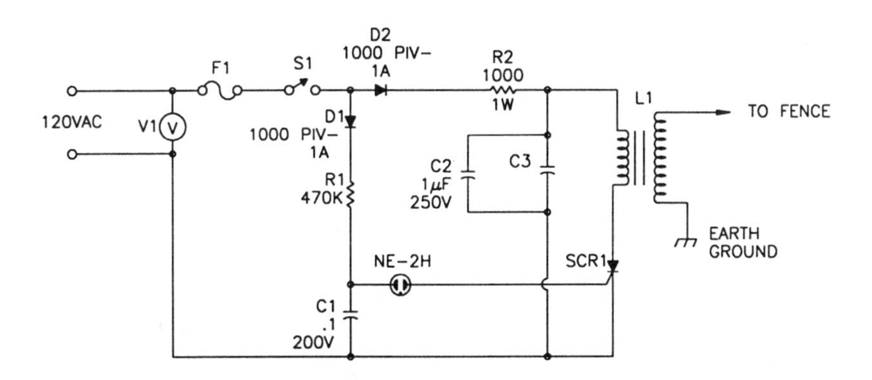

Figure 1 Complete schematic of the 'Fence Charger' circuit.

Figure 2 shows the capacitive discharge stage. It is composed of D2, R2, C2, C3, L1, and SCR1. Capacitors C2 and C3 will charge on every positive half-cycle of the AC line through diode D2 and resistor R2. Because R2 is relatively small, the capacitors will charge to nearly the peak voltage of the AC line in just one half-cycle, although they have approximately fifteen cycles to top off their charge levels. At this point, no current is flowing through the primary of L1 because SCR1 is turned off. You will recall that an SCR (silicon controlled rectifier) is normally in a high impedance state, similar to an open switch. If the anode of the device is positive with respect to the cathode, a positive pulse applied to the gate terminal will cause it to latch on, going into a low impedance state, similar to a closed switch. It will stay this way until the current flow through it ceases, or at least drops below some minimum level, known as the "holding current." If a positive pulse is applied to the gate of the SCR in this circuit (after the capacitors have charged), SCR1 will turn on, acting like a closed switch. When this happens, the capacitors will discharge through the primary winding of L1. After the capacitors discharge and the AC power line reaches the next zero-crossing, all current flow through the SCR will cease, and it will go back to its high impedance state until the next trigger pulse arrives. A spike suppression diode is not needed across the coil in the way that it is used across a relay or other inductor in a typical DC circuit. Since the SCR always turns off at a time when the current flow through the inductor is zero (or nearly so), no back EMF is generated.

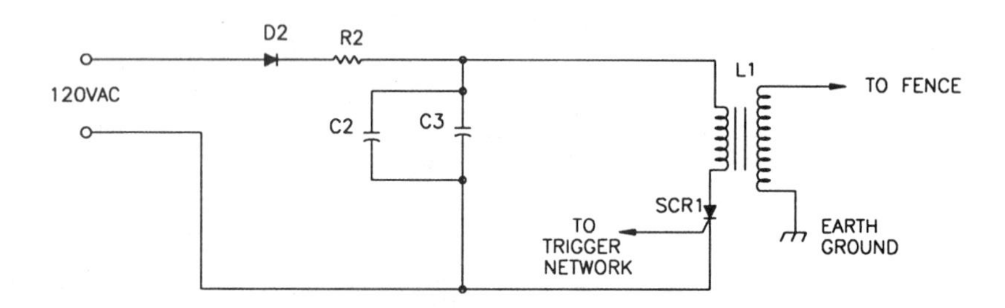

Figure 2 The capacitive discharge stage.

L1 is an ordinary 12 v coil from an automobile. It will have to be modified slightly, as described in the 'Construction Details' section. An auto coil is basically just a step-up transformer with a high-turns ratio. When the capacitors discharge through the primary winding, a high voltage spike will appear across the secondary winding. Because of the high turns ratio, the voltage output is high, but the available current is minimal, because the power coming out of the coil can be no greater than what is being applied. This allows the circuit to give a respectable jolt to anything touching the connected fence, while at current levels too low to be dangerous.

To use this circuit as a fence charger, we connect one of the secondary terminals to earth ground, and connect the other to the fence. When someone or something touches the fence, the capacitance from their body to ground completes the circuit sufficiently to give them a good jolt. In some cases, there may also be a direct resistive path to ground, which will enhance the effect. This is particularly true of animals, since they do not normally wear shoes.

The trigger network for the charger is shown in figure 3. It is composed of D1, R1, C1, and the neon bulb. This part of the circuit determines how often the coil is pulsed. On every positive half-cycle, C1 will charge to some extent through D1 and R1. When the charge on C1 reaches approximately 90-100 v, the neon bulb "breaks down" and conducts. As the gas in the neon bulb ionizes, the voltage across the bulb will abruptly drop by several volts, discharging a sufficient current through the gate of SCR1 to turn it on, which in turn pulses the coil.

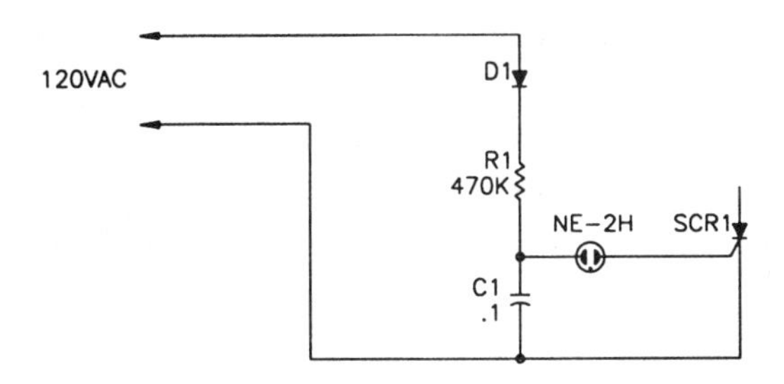

Figure 3 The trigger network.

If R1 were small enough, the trigger network could fire the discharge circuit on every positive half-cycle, allowing a maximum pulse rate of 60 Hz, but we felt that a slower rate was more appropriate, especially if the fence might be accidently touched by children. If it takes a person 1/4 second to respond to the shock and move their hand away, they might only be shocked once at a 4 Hz rate, but would get zapped fifteen times at a 60 Hz rate. We decided, therefore, to select a value for R1 that would give approximately four or five pulses per second. This would be fast enough for the application, yet slow enough to be merciful to anyone touching the fence by accident.

With a value of 470K for R1, it will generally take twelve to fifteen positive half-cycles before C1 will charge sufficiently to fire the neon lamp. This will give us the pulse rate that we desire. If you wish to alter the pulse rate for your circuit, you can increase R1 to slow it down, or decrease R1 to speed it up.

The purpose of varistor V1 (see figure 1) is to suppress any high voltage transients coming over the power line, which might damage the circuit. If this device will be attached to a fence and left on continuously, the varistor should be considered mandatory. If it will only be used briefly on rare occasions, it can be left out, although including one is always a good idea.

CONSTRUCTION DETAILS

Interior view of a completed femce charger project.

Although this circuit is relatively easy to assemble, there are a few details which must be watched due to its unique nature. Adequate insulation must be used in all wiring, since the entire circuit is at line voltage. We mounted a feed-through barrier strip on the enclosure, and connected the secondary leads of L1 to its terminals to provide connection to the fence. We recommend that you use a barrier strip of several terminals, even though only two are needed, so that the leads from L1 can be spaced at least 1" apart. Make certain to keep the secondary leads of L1 widely separated, because they are capable of arcing to one another.

Because we would not be entering the fenced area frequently in our application, we decided to omit switch S1, even though it is shown in the schematic. The mechanical nature of the switch contacts make them subject to corrosion, which can cause long-term reliability problems when exposed to a harsh environment. But if you will be entering the fenced area frequently, it may be a nuisance not to include it.

In addition to being an essential part of the circuit operation, the neon bulb can also serve as a pilot light and status indicator. The flashing of the bulb not only indicates the presence of power, but also verifies that a large portion of the circuit is operating correctly. For these reasons, we mounted it so that it would protrude through the case, rather than on the circuit board.

The neon bulb used in this project is the NE-2H. It is very important not to substitute other neon bulb types, because the current required here will exceed the ratings of most other members of the NE-2 family. Due to the less than consistent nature of the neon bulb, they tend to vary somewhat from one manufacturer to another. If after assembling the project, the triggering seems to be erratic or the neon bulb has a bluish glow when it fires, you should try a bulb from another batch or source.

Since capacitors C2 and C3 will be charging to nearly the peak voltage of the AC line, you should use ones with the highest voltage rating available. We recommend 200 v as a minimum. Units of 250 v are preferred, and 400 v units are even better. The reason we used two capacitors in parallel was simply to get the required value of capacitance from available components. A single 2 ufd capacitor will work just as well.

The output of the charger can be significantly increased by raising the value of C2 and C3. We changed the capacitance from 2 ufd to about 20 ufd, and found that we could draw an arc from the output terminals nearly twice as far as before; however, there are practical limits as to how much this capacitance can be increased. If the capacitors are too large, they will not be able to achieve a full charge through R2 in the available time between pulses. In addition, the pulse rating of the SCR may be exceeded, as well as the fact that the auto coil has certain limitations of its own which will cause the output to level off beyond a point. We do not recommend that you substitute capacitors of higher value unless you clearly need the added power output in your application. And even here, do not expect miracles--this circuit was not designed to drive a 10 mile fence around a cattle ranch.

L1 is a standard coil from an automobile. If you do not have one laying around, they are widely available at any junkyard. Although the voltage output of the charger will vary from one type coil to another, the vast majority will be fine for this project; however, it will have to be modified for operation in this circuit. The first step is to remove the coil from its case. To do this, peel back the metal casing from the plastic insulator on top of the coil (we used a large pair of pliers and a screwdriver). Then lift the coil from the casing, keeping in mind that it is probably in an oil bath. Occasionally, you may run into a coil sealed in a potting compound. If you do, you will have to discard it, unless the wires are reasonably accessible. Once you have freed the coil, you can discard the case and oil. Then clip the leads from the coil to the plastic insulator, keeping the leads on the coil as long as possible. Discard the plastic insulator. Remove and keep the steel strips from the center of the coil. Then, thoroughly clean both the coil and the strips with some type of oil-cutting solvent. We used a spray-type brake cleaner. Then allow both the coil and strips to dry for a couple of days.

Next, you will have to separate the primary and secondary ground wires, because they will generally be soldered together. This is necessary not only for the correct operation of the circuit, but also to insure total isolation between the electric fence and the AC power line. Failure to do so could cause serious injury or death. By separating the connection between the primary and secondary grounds, you eliminate the safety hazard.

When you connect the coil in the circuit, the primary winding (the one that connects to R2 and SCR1) will be the one with the larger diameter wire, usually wound on the outside. The wires from the secondary, which connect to the barrier strip (the connections going to the fence), are generally smaller in diameter. If there is any doubt at this point, a check of the resistance will reveal higher resistance in the secondary coil. Often, the secondary output terminal will be a tab, rather than a wire. If so, you can usually solder a wire directly to this tab for your connection. This lead will be your connection to the fence, and the secondary ground wire that you previously unsoldered will connect to earth ground. It is best not to connect these wires in the opposite direction, because we want the high voltage to appear on the center lead of the secondary, rather than on the outside lead. The primary winding will be relatively close to ground potential (the neutral wire is tied to earth ground), and we want to insure that the secondary does not attempt to arc to the primary winding. The polarity of the primary connection is not as critical, but we connected the former ground side of the primary to the SCR.

Before mounting the coil, you can put the steel strips back into the center. These form a core which will increase the inductance, and, therefore, improve the output of the coil. We secured the strips in the coil with silicone rubber, and then mounted the coil with this same compound. When you mount the coil, it is best if you can place it so that the secondary connections are in close proximity to their terminals on the barrier strip. Do not connect the secondary wires to two adjacent terminals because they may arc to one another.

Since this circuit involves high voltages and will normally be located outside, we recommend that a plastic enclosure be used. If you have the neon bulb protruding through the case, as we did, be sure to seal this and any other openings with epoxy or silicone rubber to keep the unit as weatherproof as possible. The box should be off of the ground sufficiently to keep it out of any standing water, and be elevated above the level of the grass, to which the hot wire can arc once it is installed and operating. Also, some shelter or covering should be provided to keep rain off of the project.

For the fence charger to work properly, it is essential that it be connected to earth ground. The best way to do this is to drive a pipe or ground rod into the ground a couple of feet, and use a clamp to fasten the wire from the charger to the pipe. For testing, the ground connection on an electrical outlet can be used, but we do not recommend this for a permanent installation.

To put up an electric fence, you will need the wire, insulators, and stakes or fenceposts to support it. Many hardware stores will carry these items. Almost any bare wire of suitable gauge can be used if you have some available, but do not try to improvise when it comes to the insulators. These provide the very high resistance that we need, and are generally so reasonably priced that it is a waste of time to search for a substitute. Once you have the fence erected and connected to the circuit, check to make sure there are no arcs to the posts. Keep any grass or weeds mowed under the fence, since it may arc to them if they get too tall.

PARTS LIST

Semiconductors:

D1, D2- 1000 PIV-1 amp (Radio Shack #276-1114)
SCR1- 400 v-6 amp (Radio Shack #276-1020)

Capacitors:

C1- .1 ufd-200 v (Radio Shack #272-1053)
C2, C3- 1 ufd-250 v (Radio Shack #272-1055)

Resistors:

R1- 470K-1/4 w resistor (Radio Shack #271-1354)
R2- 1000 ohm-1 w resistor (Radio Shack #271-153)

Miscellaneous:

F1- 1/2 amp fuse and holder
L1- auto coil
NE- 2H neon bulb (Radio Shack #272-1102)
S1- 120 v-1 amp SPST switch
V1- varistor (Radio Shack #276-570)
Printed circuit board (Radio Shack #276-150)
Project enclosure (we used a Radio Shack #270-224)
Feed-through barrier strip, power cord, PC board standoffs, rubber feet for case, misc. hardware

BRINGING UP THE UNIT

Once the fence charger has been completed, a continuity check should be performed for safety, just to verify that the AC line has been correctly isolated from the fence. With the unit unplugged and the switch turned on, use an ohmmeter to check between each lead of the power plug and both output terminals. Then reverse the leads and measure again. There should be no continuity read on either measurement.

After the isolation has been verified, plug the charger in and turn on the switch. The neon bulb should immediately start to flash (at about a 4 Hz rate), and you should be able to hear the coil pulsing. If everything seems to be functioning correctly, you can test the output by attaching a jumper lead to the ground output terminal and bringing the other end of the jumper near the hot output terminal. You should be able to draw a small arc. If you decide to test it instead by simply touching the output terminal (as all real men do), you will probably feel only a very slight shock at this point, since we do not have the earth ground connected yet. Keep in mind that capacitors C1, C2, and C3 have now been charged, and may be a potential shock hazard even after power has been removed. Always short these components out before servicing the unit.

If the unit does not seem to function properly, check the following:

- Verify that the neon bulb is flashing. If so, then the trigger circuit is functioning, and the problem resides in the capacitive discharge stage (see step 2 below). If the bulb is not flashing, then the problem is in the trigger portion of the circuit. Looking at figure 1, begin troubleshooting by taking a voltage measurement from the bottom of C1 to the top of D1. You should read about 120 v. If not, check fuse F1, and for the proper operation of switch S1. If the power is present, measure the DC voltage across C1. If the voltage is above 120 v, then you either have a bad neon bulb or SCR. To verify which is at fault, short the gate and cathode of the SCR. If the bulb fires now, the SCR is at fault; otherwise, the neon bulb should be replaced. If the voltage on C1 is low, however, you probably have a bad capacitor. Although less likely, D1 or R1 could also cause this problem. Make sure that D1 is not installed backwards.

  If the unit functions, but the triggering seems erratic, or the neon bulb has an odd color to it when it flashes, remember to try a neon bulb from another source or batch.

- Once you have verified proper operation of the trigger circuit, we can proceed to troubleshoot the capacitive discharge stage. Start by measuring the voltage across the SCR. If the voltage here consistently stays above 120 v, and the gate of the SCR is being triggered, replace the SCR. If the voltage across the SCR is very low or zero, temporarily disconnect one of the leads from the primary winding of L1. Then measure the voltage across C2 and C3. If the reading is about 170 v or so, either the SCR is shorted, or the primary of L1 is open. If the voltage across these capacitors is very low or zero, D2 or R2 could be open, or else C2 or C3 is shorted. Make certain that D2 is not installed backwards.

LONG-TERM LINEAR TIMER

The need often arises for a simple timer circuit. For time delays of up to an hour or so, the 555 IC is a popular choice. For very long delays, or very precise timing intervals, a digital timer must generally be built. Occasionally, however, an application comes along that requires a timing interval in the 30 minute to several hour range. While the 555 is capable of doing this, the large values required for the timing resistor and capacitor can make the design difficult. The large electrolytic capacitors almost always have a lot of leakage. These high-valued components can also be difficult to locate and obtain. If you want variable control over the timer's range and need a potentiometer of very high resistance, the value you need may not even be available. Although digital methods can always be employed in these situations, this much accuracy is not always required for the hobbyist's application, and it would be nice if we could simply use a 555-type IC to generate an analog delay for this long of a period, without resorting to the use of oddball or insanely valued components.

There have been several ICs geared to this application in recent years. They almost always incorporate an oscillator followed by a counter/divider stage, which divides the oscillator frequency down to obtain the required period; however, these chips are not really mainstream, and they are not always easy to obtain. Since most commercial and industrial applications generally would not use an analog delay of this length, these ICs often fall by the wayside. If you build a device around one of them, and later need a replacement, it may no longer be available. Another technique is to use a 555 timer in an astable mode to generate pulses, and then connect a counter to the output, but the output of the counter usually cannot power a relay directly, so an output device is also required. Then, some gating circuitry must be included to stop the device after time-out, or else it will keep repeating its cycle. By the time you add all of this in, you may as well go ahead and build a digital timer.

The circuit presented here is a relatively simple analog timer using one IC (not counting the power supply regulator), the popular 556, which includes two 555 timers in one package. This circuit can be used for timing periods of several minutes, on up to several hours. It requires no ridiculously high part values, nor the use of any oddball components; all parts are mainstream and readily available. As an example, a typical 555 timer circuit would require a potentiometer of at least 33 meg to generate a 1 hour delay if a timing capacitor of 100 ufd was being used. We will show you how a 1 meg potentiometer can be used with this same capacitor to generate a delay of the same length. Accuracy of this circuit is typical for an analog delay of this type; generally, it will time out within plus or minus of 1 minute per hour of delay. Although this obviously will not match digital accuracy, it is sufficient for many hobbyist applications, and its simple one IC design makes it fast and easy to assemble.

Exterior view of completed long-term linear timer project.

CIRCUIT DESCRIPTION

Before we begin describing this circuit, let's take a look at a typical 555 timer, and see why we tend to run into problems when we attempt to use it for long delays. If you are not familiar with the internal structure of the 555 IC, you may wish to read the portion of the chapter on the 'Light Sequencer' project concerning it. There, we explain the essentials of what is contained inside of it.

Looking at figure 1, we see a typical circuit of this type. Power is supplied through pins 1 and 8, and pin 4, the 'reset' terminal, is active when low; therefore, we use a resistor to hold it high, unless the 'reset' button is pressed. Pin 5, the 'control' terminal, is used for varying the trigger and threshold points. It is not used in this application, so we leave it unconnected, except for a bypass capacitor to give us the best possible stability. Pin 3, the 'output' terminal, is normally low. When the circuit is triggered, it will go high, turning on the relay, until the time delay has expired. Pin 2, the 'trigger' terminal, starts the timing operation. It will begin the delay when the voltage on it drops below one-third of the power supply voltage; therefore, we use a resistor to hold this pin high until the 'start' button is pressed, which pulls the terminal low to start the delay. Before the delay starts, a discharge transistor connected to pin 7 is turned on, holding the timing capacitor, Ct, at ground potential. When the time delay starts, however, this transistor is turned off, and Ct starts to charge through the timing resistor, Rt. When the voltage on Ct reaches two-thirds of the supply voltage, the time delay ends. Pin 3 goes low, turning off the relay, and the discharge transistor on pin 7 turns on to quickly discharge the timing capacitor in order to prepare for the next cycle.

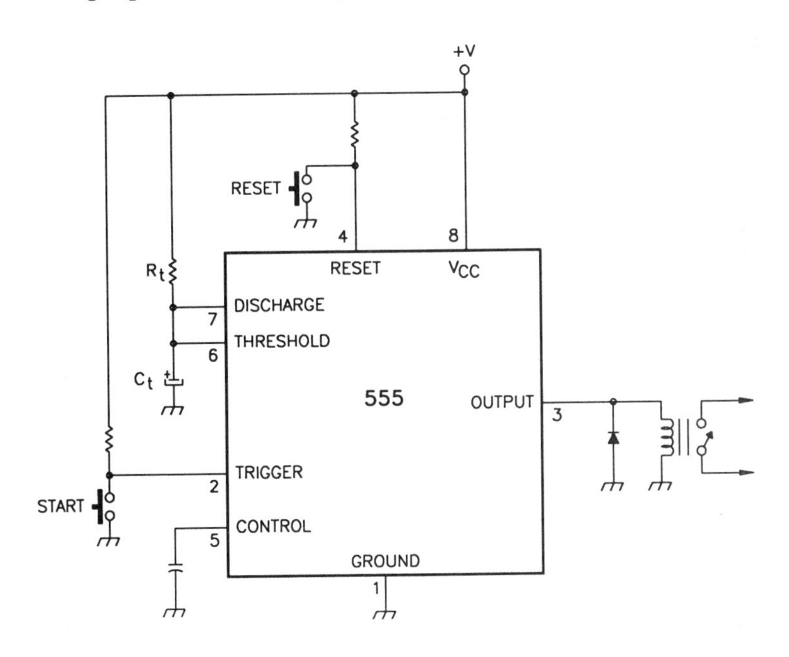

Figure 1 A typical 555 timer circuit.

The 555 was a well thought-out design, and it makes an excellent analog timer, generally having an accuracy of plus or minus a second or so per minute of delay, possibly a little better with premium timing components. Although it is capable of generating a time delay of hours, it is often difficult to achieve this because of the large electrolytic capacitors required. These capacitors tend to have a considerable amount of leakage, and the larger the capacitor, the greater the problem, since the amount of leakage is proportional to the area of the capacitor's plates. Generally, you will start running into leakage problems with capacitors of 1000 ufd and above, although the lower-valued units can also cause occasional problems.

A perfect capacitor should not allow any DC current flow through it. Although no capacitor is perfect, most types allow only a very small amount of leakage through their insulator. The electrolytic capacitor, however, forms its insulating layer through an electrochemical process. This allows a large amount of capacitance to be put in a small case, yet also allows a larger DC current flow through its dielectric. This leakage current is still small enough that it does not cause a problem in most applications, but when a large resistance is placed in series with it, this current flow will result in a voltage drop across the series resistor, limiting the voltage to which the capacitor can charge. Looking at figure 2, we can think of this DC leakage as a resistor in parallel with the capacitor, effectively forming a voltage divider. With a series resistor of 1 meg and a leakage current which causes the capacitor to appear as though it had a 1 meg resistor in parallel with it, the capacitor will not charge to more than 1/2 Vcc, since the voltage divider set up by the two resistors will stabilize at this point. If this particular resistor-capacitor (RC) combination was connected to a 555 timer, the delay would never time out, since the capacitor would never reach 2/3 Vcc.

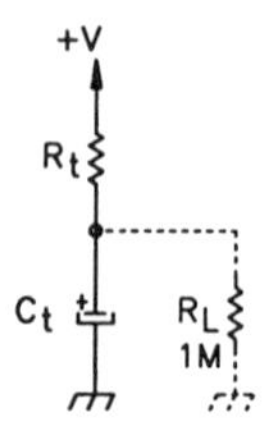

Figure 2 The DC leakage of a capacitor can be thought of as a resistor in parallel, forming a voltage divider with Rt, although this model is somewhat simplistic when dealing with electrolytic capacitors because their insulation resistance is voltage-sensitive due to their electrochemical construction; therefore, the actual value of the leakage resistance would be constantly changing in the circuit above.

A more common situation is to find that your capacitor will eventually charge to 2/3 Vcc, but the leakage current has been sufficiently draining off the charge to greatly prolong the delay. If the leakage current is such that it allows a capacitor to charge only up to 9 v in a 12 v circuit, the voltage across it will eventually reach the 8 v level needed to reset a 555, but the 2 hour delay you were expecting may run for 5 hours.

Often, you will have better luck if you try using a smaller capacitor in combination with a large resistance value. While the standard 555 timer is made to operate with a timing resistor of up to 10 meg, higher values can generally be used, and the CMOS version can use values up to 100 meg, although its output sourcing current is rather meager if you need to drive a relay. The disadvantage of this approach is that it makes variable control over the entire range very difficult. If you will always be using a timer for a 1 hour delay, you can stack three 10 meg resistors in series with a 5 meg potentiometer, and use a 100 ufd capacitor. But what if you need to be able to adjust the timer for different delays throughout its range? You then have the less than enviable task of trying to locate a 35 meg potentiometer. Since it can be frustrating to try to find and use resistors, potentiometers, and capacitors of such extreme values, it would be nice if we could modify the circuit in some way to allow the use of readily available components. Is there another approach that we can take to solve this problem?

Looking again at figure 1, let's assume that Rt has a value of 1 meg and Ct has a value of 100 ufd. According to the formula used for calculating one-shot times for the 555 (T = 1.1 RC), this should give a delay of about 110 seconds, or 1 minute and 50 seconds. Now imagine that we have placed a switch in series with Rt, and that the switch is closed. When the 555 is triggered, Ct will start to charge. After 1 second, however, we open the switch. We wait 1 second, and close it again. One second later, we open it again, and so on, until the charge on the timing capacitor reaches 2/3 Vcc and the circuit times out. How long would the delay last?

Since the capacitor was charging only half of the time, the delay would be twice as long as before, or 220 seconds. We have doubled the time delay using the exact same timing elements.

Now suppose that we trigger the circuit again, but this time we close the switch for 1 second and leave it open for 9 seconds. Since the capacitor is only charging one-tenth of the time, the delay has been extended by a factor of ten times, still using the exact same timing elements. Our 1 minute and 50 second delay has now been extended to over 18 minutes.

Of course, to make this concept practical, we must find some other way to switch the resistor in and out of the circuit. No one wants to use a timer that requires them to manually open and close a switch every few seconds; that is where the other half of the 556 we are using comes in.

Figure 3(a) shows a typical 555 astable circuit. This will function as a squarewave oscillator, with the output going high and low at regular intervals. If we disconnect the timing resistor from the one-shot in figure 1 from Vcc, and connect it to the output of this oscillator through a diode (to prevent Ct from discharging through Rt on the low part of the cycle), then Ct will be receiving a charge only for the percentage of time that the oscillator's output is high. If we design the oscillator to have a waveform where the high portion of the output is relatively short compared to the low time, we will greatly extend the delay period of our timer.

Before we can do this, however, a refinement to the oscillator in figure 3(a) is needed. You will recall that when power is applied to this circuit, the output will go high, since capacitor C initially holds the trigger terminal low. As the capacitor charges through resistors Ra and Rb, the voltage on C will eventually reach 2/3 Vcc. The output then switches low, and the discharge transistor turns on, effectively bringing the 'discharge' pin to ground. Then C discharges through Rb until 1/3 Vcc is reached, at which point the cycle repeats. The problem here is that during the time the output is high, C charges through both Ra and Rb, but during the low time, C discharges only through Rb. This means that the output must always be high for a longer period of time than it is low; thus, we cannot use this circuit to generate a waveform that is high for only a small percentage of the time.

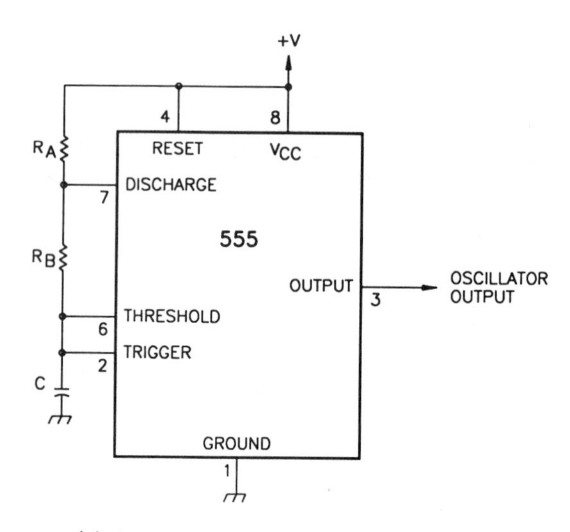

(a) A typical 555 astable multivibrator.

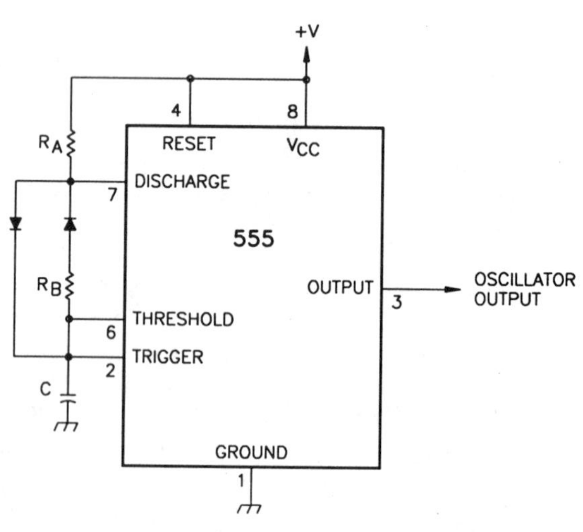

(b) 555 astable circuit with "steering" diodes added.

Figure 3

This ratio of 'on' time to 'off' time is known as the "duty cycle." Sometimes it is expressed as a percentage. A waveform that is high for one time interval and low for nine time intervals has a duty cycle ratio of 1:9. Since the waveform is high one-tenth of the time, it could also be expressed as a 10% duty cycle. For our application, we need a relatively low duty cycle, percentage-wise. A 10% duty cycle would cause our time delay to extend by ten times, a 5% duty cycle would extend the delay by twenty times, but a duty cycle of 50% would only double the delay time, and the oscillator of figure 3(a) cannot even give us this. Since the output of that circuit must always be high for a longer period of time than it is low, it will always produce a duty cycle above 50%.

When low duty cycles are needed with a 555 astable circuit, the most common solution is to add a pair of "steering" diodes. Figure 3(b) shows how they are connected. What this does is give us independent charge and discharge paths. Now, C will charge only through Ra, and discharge only through Rb. This allows us to select any duty cycle that we want.

Figure 4 shows the synthesized circuit, with the output of the oscillator connected to the timing resistor of the timer stage through diode D3, and a power supply added. If this circuit is starting to look a little involved, keep in mind that this is all based on one IC. We show the circuit as separate stages for clarity , but the construction time for this circuit is probably only 10-15 minutes longer than that for the traditional 555 timer of figure 1.

In our prototype, we used a resistance of 4.7K for Ra (R1), and 133K for Rb (R2+R3). This will produce a duty cycle of about 3.4%. Since the resistor ratio of Rb to Ra is 133/4.7, or 28.3, the output will be low 28.3 time intervals for every one time interval that it is high; thus, the output will be high 1/29.3 of the time, and our time delay should be extended by a factor of 29.3.

The "multiplication factor" of the time delay can be expressed by the formula (Rb/Ra)+1, or (Rb+Ra)/Ra, and the total delay period can be expressed by the formula (Rb/Ra+1) (1.1 Rt Ct). For the part designations in figure 4, this formula would be T= ((R2+R3/R1)+1)(1.1 (R7+R8) C5). With the values shown for our prototype, this should give a maximum time delay of about an hour. In practice, however, we found that the delay generally exceeds this by 5-20%. Part of the reason for this is the fact that even when the output of the oscillator stage is high, R7 is not connected to full Vcc voltage; there is a drop across diode D3. There is also some capacitor leakage involved, as well as a wide variation in the actual amount of the capacitance of C5. As a general rule, when using the circuit of figure 4, you will obtain a little over 1 hour's delay for every 100 ufd of capacitance for C5.

Figure 4 also shows the power supply for the circuit. Power is applied through fuse F1 and switch S1, both of which must be sufficient to carry the current of the controlled device. The purpose of fuse F2 is to protect the circuitry itself, and varistor V1 protects the circuit from line transients. Transformer T1 reduces the line voltage to 12.6 vac, which is rectified by bridge B1, and filtered by capacitors C7 and C8, and the 7812 regulator maintains a steady output voltage of +12 v. Capacitor C9 improves the transient response of the regulator, while C10 and C2 provide RF bypassing for the supply (capacitor C2 should be placed as close to the 556 as possible). Resistor R9 insures a minimum current flow through the regulator for proper operation. With the circuit as drawn in figure 4, 120 v is supplied to the load when relay K1 is energized. An optional 120 v pilot light is included to indicate that power is applied to the load.

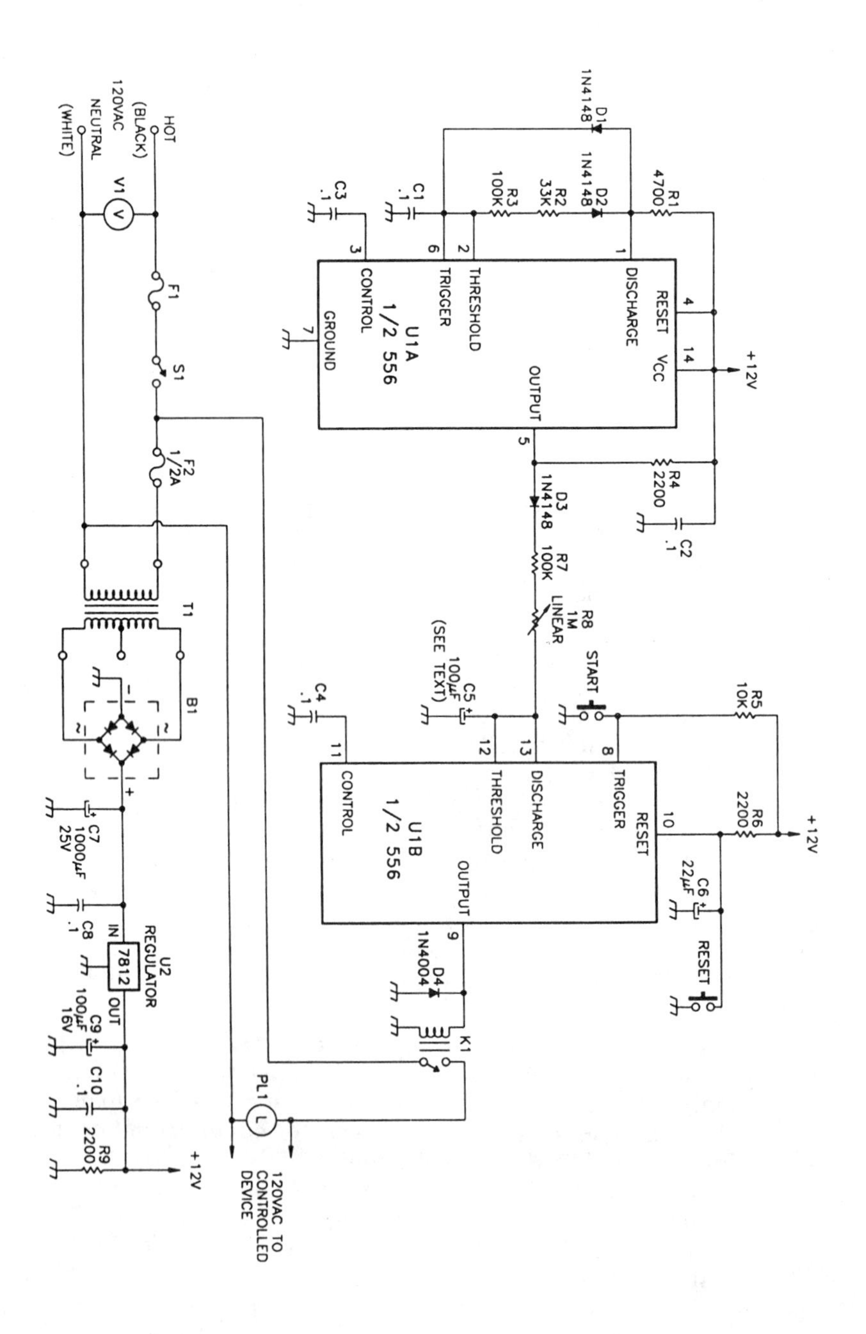

Figure 4 The synthesized timer circuit.

Although the circuit in figure 4 does everything that we need it to do, we decided to modify the prototype to turn off the circuit, as well as the load at the end of the timing cycle. This is a good idea for circuits that will not be frequently used. Figure 5 shows the final circuit for the prototype. To do this, we started by applying power to both the circuit and the load through the relay contacts. We moved the 'start' switch so that it shorted the relay contacts. When the switch is pressed, power is applied to the transformer. When the circuit powers up, the relay must pull in to keep the circuit powered; therefore, we have moved capacitor C6 from the 'reset' terminal of the 556 to the 'trigger' terminal, so that the timer will trigger on power-up. This guarantees that K1 will pull in, holding the circuit on even after the 'start' button has been released. At the end of the time delay, K1 will de-energize, turning off both the load and the timer circuit.

Although the circuit in figure 5 may be a better choice if the timer will not be used frequently, it has one inherent disadvantage, as well. When the delay times out, the 'discharge' terminal (pin 13) goes to ground, discharging C5; however, due to the high internal series resistance of an electrolytic capacitor, it takes a certain amount of time to completely discharge C5. Since this circuit also cuts itself off at the end of the delay, it is likely that a small residual charge will remain on the timing capacitor after the circuit has died. This will drain off in time, due to leakage through the various components, but if a second delay is started shortly after the first delay has expired, the second delay may be slightly shorter than expected, since the timing capacitor had a "head start." If this will be a problem, but you really prefer the automatic circuit cut-off of figure 5, one simple solution is to substitute a DPDT relay for K1, and use the second set of contacts (NC) to short out C5 when the relay is de-energized. This will guarantee that every timing cycle starts out with C5 completely discharged.

This technique of switching a component in and out of the circuit to make it appear as though it were a different value has many other applications. Such circuits, known as "duty-cycle integrators," can be used to make light dimmers for LEDs and DC incandescent lamps, and DC motor speed controls, and are also the basis for "switching"-type power supplies. For example, an LED attached across the output of the oscillator stage of this circuit (with a suitable current limiting resistor) will glow brighter as you increase the percentage of time that the output is high. By using the oscillator to turn a power transistor on and off, heavier DC loads can be controlled, such as small motors. Using a potentiometer to change the duty cycle will vary the amount of power delivered to the load. This technique is much more efficient than using a series resistance to control the lamp intensity or motor speed because the switching device is always turned either fully on or fully off; therefore, the power dissipated by the controlling device is virtually zero, whereas a series resistance can often dissipate as much power as the load itself.

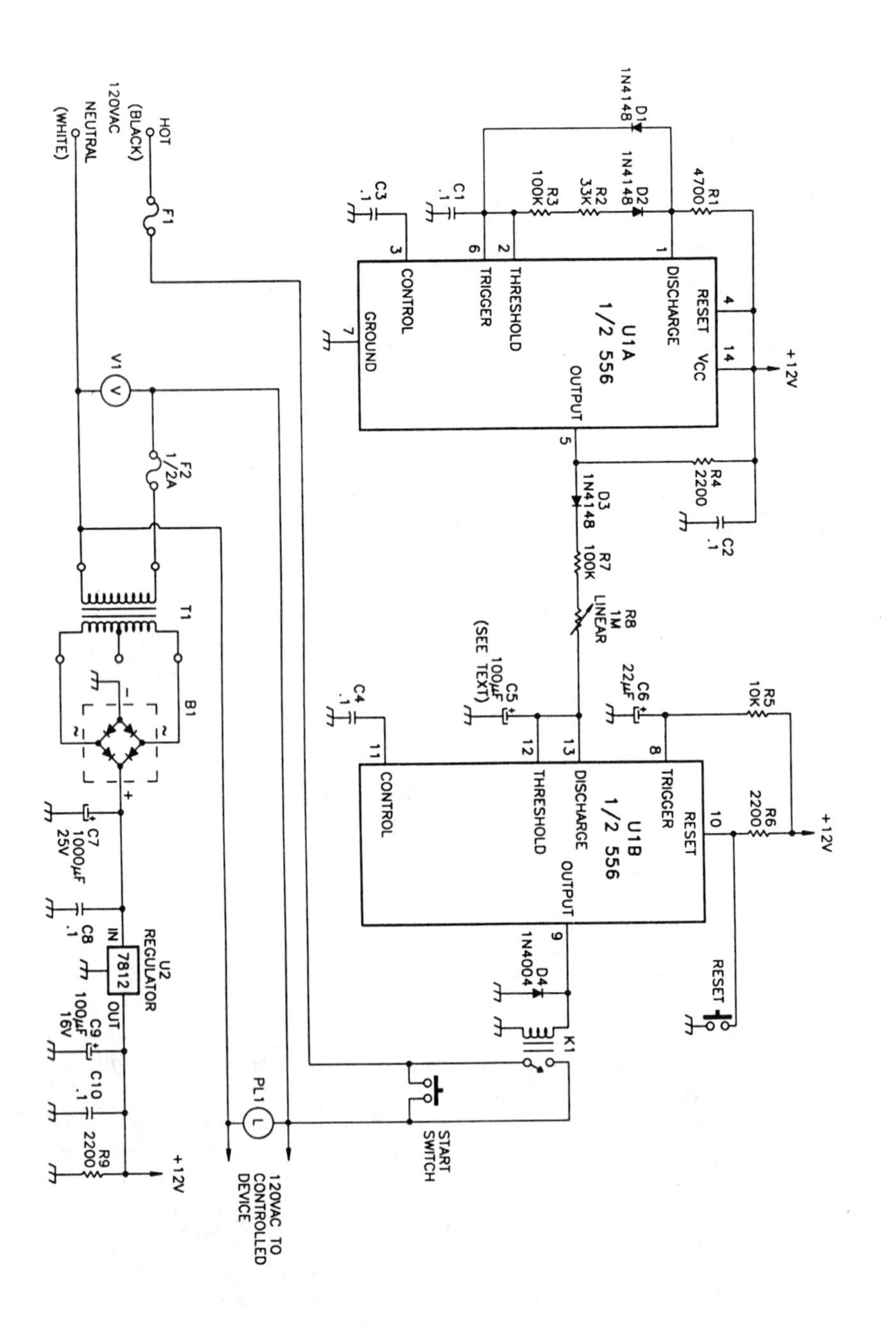

Figure 5 Circuit modified to cut itself off.

CONSTRUCTION DETAILS

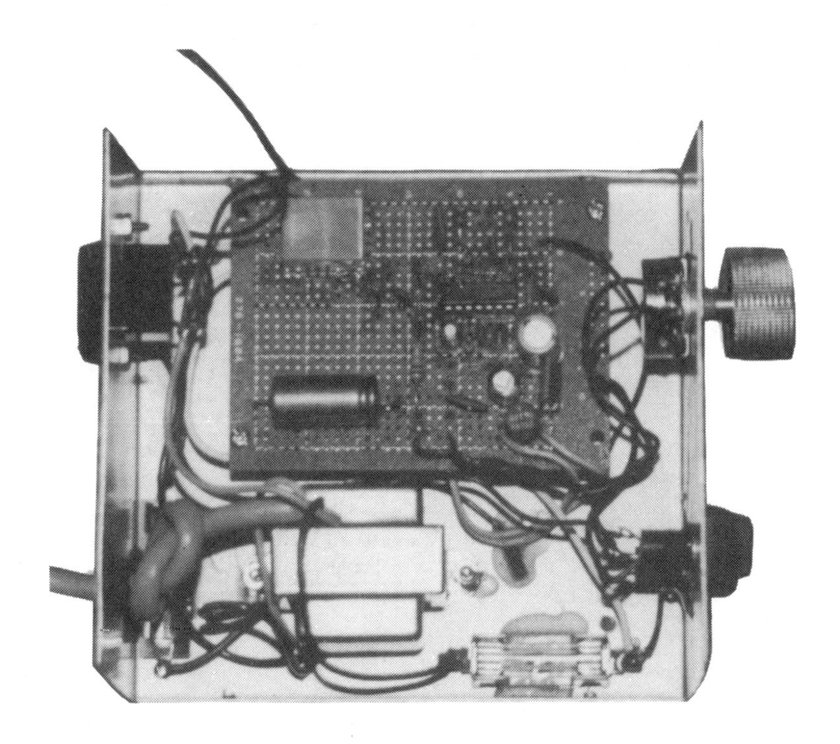

Interior view of a completed long-term linear timer project.

Construction of this project is rather straightforward, and is similar to that of any linear timer. We built our prototype on a Radio Shack board (#276-168), which had plenty of room for all the components. The case we used was also from Radio Shack (#270-253), but just about any project box will work. If you use a metal case, as we did, use a three-wire power cord so that you can attach the ground wire (green) to it for safety.

If you mount the relay on the circuit board, and it will be switching a 120 v load, make certain to isolate the high voltage lines from the low voltage circuit.

The values shown in our prototype will generate a delay of about 6 minutes to over an hour, depending upon the setting of the potentiometer, R8. If you wish to change the range of delays, you have the option of either changing the value of the timing capacitor, or of altering the duty cycle by changing R1, R2, and R3. With the value these resistors have in figures 4 and 5, you will get a little over one hour's delay for every 100 ufd of capacitance for C5. The time delay range can also be altered by changing the duty cycle; however, there will be practical limits as to how far you can go. In the prototype, the oscillator output is high only about 3.4% of the time. You will probably not want to go much farther than this, because the capacitor leakage will become a dominant influence, possibly draining off the charge faster than it is being pumped in. Of course, shorter delays are not a problem.

You can also substitute a 100K trimmer potentiometer for R3. Adjusting this will allow you to change the range of R8, or calibrate the circuit so that R8's position of maximum resistance will correspond to some particular length of time.

In this circuit, we are pushing the timing capacitors to their physical limits. We are reaching a point where their leakage must be minimized for best operation. New electrolytic capacitors, or ones that have not been used for extended periods, may tend to exhibit higher leakage than normal. This is due to deterioration of the dielectric layer. This can usually be restored by connecting the capacitor to a voltage source, and allowing it to reform the insulating layer. For this circuit, it might be a good idea to 'exercise' the timing capacitor before installing it by connecting it to a voltage source for a few hours. Or you can simply install it in the circuit and let the capacitor build up its dielectric while in operation. The first few cycles will probably be erratic, and the first delay may be considerably longer than expected. Regardless, it is a good idea to run at least five cycles, with R8 at its maximum resistance, before trying to adjust the time. Keep in mind that if the timer is not used for an extended period, the capacitor's dielectric may once again deteriorate, causing the first cycle to be slightly longer than expected.

PARTS LIST

Semiconductors:

B1- 100 v-1.4 amp bridge rectifier (Radio Shack #276-1152)
D1, D2, D3- 1N4148 silicon diode (Radio Shack #276-1122)
D4- 1N4004 silicon rectifier (Radio Shack #276-1103)
U1- 556 dual timer IC (Radio Shack #276-1728)
U2- 7812 +12 v regulator IC (Radio Shack #276-1771)

Capacitors:

C1-C4, C8, C10- .1 ufd-50 v disc
C5- (see text - we used a 100 ufd electrolytic for a 1 hour delay)
C6- 22 ufd-16 v electrolytic
C7- 1000 ufd-25 v electrolytic
C9- 100 ufd-16 v electrolytic

Resistors: (All resistors 1/4 w unless stated otherwise)

R1- 4.7K
R2- 33K
R3- 100K
R4, R6, R9- 2.2K

R5- 10K
R7- 100K
R8- 1 meg potentiometer-linear taper

Miscellaneous:

F1- fuse and holder (value of fuse will depend on load, but must be less than 10 amp)
F2- 1/2 amp fuse and holder
K1- 12 v relay (Radio Shack #275-248)
S1, S2- normally open push button switches-120 v (we used Radio Shack #275-1566)
S3- 120 v toggle switch **Optional (for use in circuit of figure 4 only)
T1- 12.6 v-450 mA power transformer (Radio Shack #273-1365)
V1- varistor (Radio Shack #276-570)
Circuit board (we used a Radio Shack #276-168)
Project case (we used a Radio Shack #270-253 metal cabinet)
Line cord, circuit board standoffs, misc. hardware
**Optional 120 v pilot light (PL1) - (Radio Shack #272-712)

BRINGING UP THE UNIT

Operation of the timer circuit is straightforward and self-explanatory. Pressing the 'start' button starts the timing cycle, and pressing 'reset' will abort the cycle. Keep in mind that if the timing capacitor is new, or has not been used for some time, the first couple of cycles may run longer than expected, since this circuit requires a minimum of leakage to work properly.

If the unit does not seem to function properly, check the following:

- Verify that the power supply voltage across the 556 is correct. In measuring from pin 14(+) to pin 7(-), there should be very close to 12 v. If there is, proceed to step 2 below. If not, measure the voltage on the input of the regulator. If it is correct (15-18 v), but the voltage coming out is wrong, first make certain that you have not accidently shorted Vcc to ground, then replace the regulator, if necessary. If the voltage coming into the regulator is low or zero, double check the wiring of the transformer and bridge, and make certain that both fuses are good. A short or solder bridge from Vcc to ground may also pull the voltage down on the regulator input. A short on the output side of the regulator will not necessarily blow the fuse, because the regulator has built-in short circuit protection.
- If the relay does not energize when you press the 'start' button, go to step 3 below. If the relay pulls in correctly, but the time delay does not end, measure the voltage on pin 5 to ground. This is the output of the oscillator. If it is working correctly, the voltage you measure should be

12 v multiplied by the fraction of time that the output is high. For example, with a 50% duty cycle, the output is high one-half of the time. You should, therefore, get a reading of 12 x 1/2 = 6 v. With the low duty cycle used in the prototype, you should read about 12 x 1/29.3 = .41 v. If you read 0 or 12 v on pin 5, the oscillator is not working. Check the wiring of this stage carefully, and if necessary, replace the IC. If the oscillator is working, but the delay does not time out, temporarily short the cathode of D3 to +12 v, and press the 'start' button. In doing this, you are making the circuit function just like the timer in figure 1. The time delay should now last for a period of T= 1.1RC, where T is the time in seconds, R is the value of the timing resistor in megohms, and C is the value of the timing capacitor in microfarads. For the values shown in the prototype of figure 4 or 5, R would consist of R7 and R8 (1.1 meg), and C would be the value of C5 (100 ufd). The time delay, with the potentiometer in the position of maximum resistance, would equal 1.1 ([1.1 megs][100 ufd]=121 seconds), or 2 minutes and 1 second. If the delay times out correctly now, the timing capacitor probably has too much leakage. You can either try another capacitor, or try "exercising" the present one. Often a new electrolytic will have a considerable amount of leakage until it has had a voltage applied to it for a time, giving it a chance to renew its dielectric layer. You can either remove it from the circuit and connect it to a voltage with a value near its working voltage for a few hours, or else you can leave it in the circuit and let it slowly develop its insulating layer while in operation. Simply trigger the circuit and let it take as long as necessary to reset the 556. Repeat until the timing cycles shorten to the correct time and gain a reasonable amount of consistency. Large electrolytics may have too much leakage to ever reset the 556, but we tested many capacitors in the 100-1000 ufd class, and all of them worked.

If shorting the cathode of D3 to +12 v as described above still does not make the timer reset in a reasonable time, check the voltage across C5 with the circuit still triggered. If, after the delay, the voltage is well below what it should be, either C5 or the 556 IC is bad. If the voltage has exceeded 8 v, but the time delay has not ended, make certain that the voltage on pin 8 is near +12 v. If it is, replace the 556 IC.

- If the power supply voltages are correct, but the relay does not pull in when the 'start' button is pressed, check the voltage on pin 10 (the 'reset' terminal). If you have built the circuit in figure 5, you will have to hold the 'start' button in to keep power applied to the circuit. The voltage here should be around +12 v. If this is correct, verify that the voltage on pin 8 really is going low to 0 v when the 'start' button is pressed (figure 4), or that the capacitor on pin 8 is momentarily holding the terminal low on power-up (figure 5). If shorting pin 8 to ground triggers the timer, replace the capacitor. If everything here is correct, check the voltage on pin 12. The voltage here should be very low at this point, close to 0 v. If it is over 8 v, check the wiring carefully around C5 and R8. If a problem cannot be found, replace the IC.

- If the circuit should ever show a tendency to re-trigger when it cuts off at the end of a timing cycle, it is probably due to transients being generated by the relay contacts arcing. Placing a varistor similar to V1 across the contacts should take care of this. An additional .1 ufd disc capacitor can be placed across the power supply lines near the 556, if necessary. Placing a .1 ufd disc capacitor across R5 (for the circuit in figure 4) may also help. The noise immunity of this circuit is quite large, however, and it is not likely that you will have a problem of this nature.

WATER REPLENISHMENT SYSTEM

This project is a water replenishment system that is designed to maintain the water level in a container. The container could be anything from a birdbath to a hog trough. Our application for this device was to automatically refill a dog's water bowl. Although the most obvious use for this project is the hands-off watering of animals, there are also many other uses, such as maintaining the water level in fountains and pools.

This circuit uses a small electronic sensor plate to determine the need for water. When the water drops below the sensor, the system will turn on to refill the bowl to a predetermined level. Although mechanical float valves are commercially available, this system has the advantages of being able to operate in areas of limited space, and also of greater aesthetic appeal. In many applications, both of these factors are important. For instance, consider a system to maintain the water level in a birdbath. Even if there was adequate space for a mechanical valve, not many people would want what resembles the inside of a commode prominently displayed in their front yard. By contrast, all that this system would require is a small wire for the sensor and a length of aquarium hose to deliver the water.

CIRCUIT DESCRIPTION

Before attempting to design a device such as this, we must clearly define exactly what is expected of it, and then outline a logical method of achieving what we are after. Obviously, we will need a detector circuit of some sort to indicate whether water is present or absent. Our first inclination might be to simply allow this detector to directly drive our electric water valve, turning on the water whenever the sensor is dry, and cutting it off when it is wet. Although it is not uncommon to see fluid detector circuits in project books laid out in this manner, a moment's reflection will reveal that such poor designs have not been thought out very well, and probably were never really built and tested.

A simple detector circuit is sufficient for lighting an LED or indicator lamp, but in a closed loop system such as we are designing, we need to introduce some hysteresis so that when the water turns on, it will go ahead and run for at least a reasonable amount of time, stopping at a point higher than its cut-on point. If we do not, the device may be turning on and off a ridiculous number of times per day, since every time it turns on it will only be able to run for a second or two before water once again makes contact with the sensor. In larger bodies of water, it may even create oscillation problems. When the water drops a small fraction of an inch below the sensor, the water will turn on. If it creates even a minute wave pattern, the valve will be turned on and off every time a wave comes by, until the water level is sufficiently high to turn it off completely. Our design should be free of this, positively turning the water on and off without any hesitation or oscillation.

Exterior view of a water replenishment system as used for an automatic dog watering container.

There are a number of ways in which we can overcome this problem. One obvious method would be to use two sensors and two detectors. One sensor would be used to turn the water on, and the other would turn it off. By placing the turn-off sensor somewhat higher than the other, we can prevent any of the problems just discussed. The obvious disadvantage of this system is that you need two sensors and two detector stages. Another method, and the one we chose, is to have a re-triggerable time delay activated when contact with the water is established. When the sensor has made firm contact with the water, a time delay will start. When it expires, the water valve is turned off. The water level will always stop at the same point, since the delay cannot begin until the sensor is submerged. This, too, will solve the problems mentioned before, but now we will need only one sensor.

The complete schematic for the watering system that we built is shown in figure 1. The circuit consists of a detector stage, which detects whether the water is present or not, and a timer stage, which is a re-triggerable time delay built around a 555 IC. Of course, we have also included a power supply.

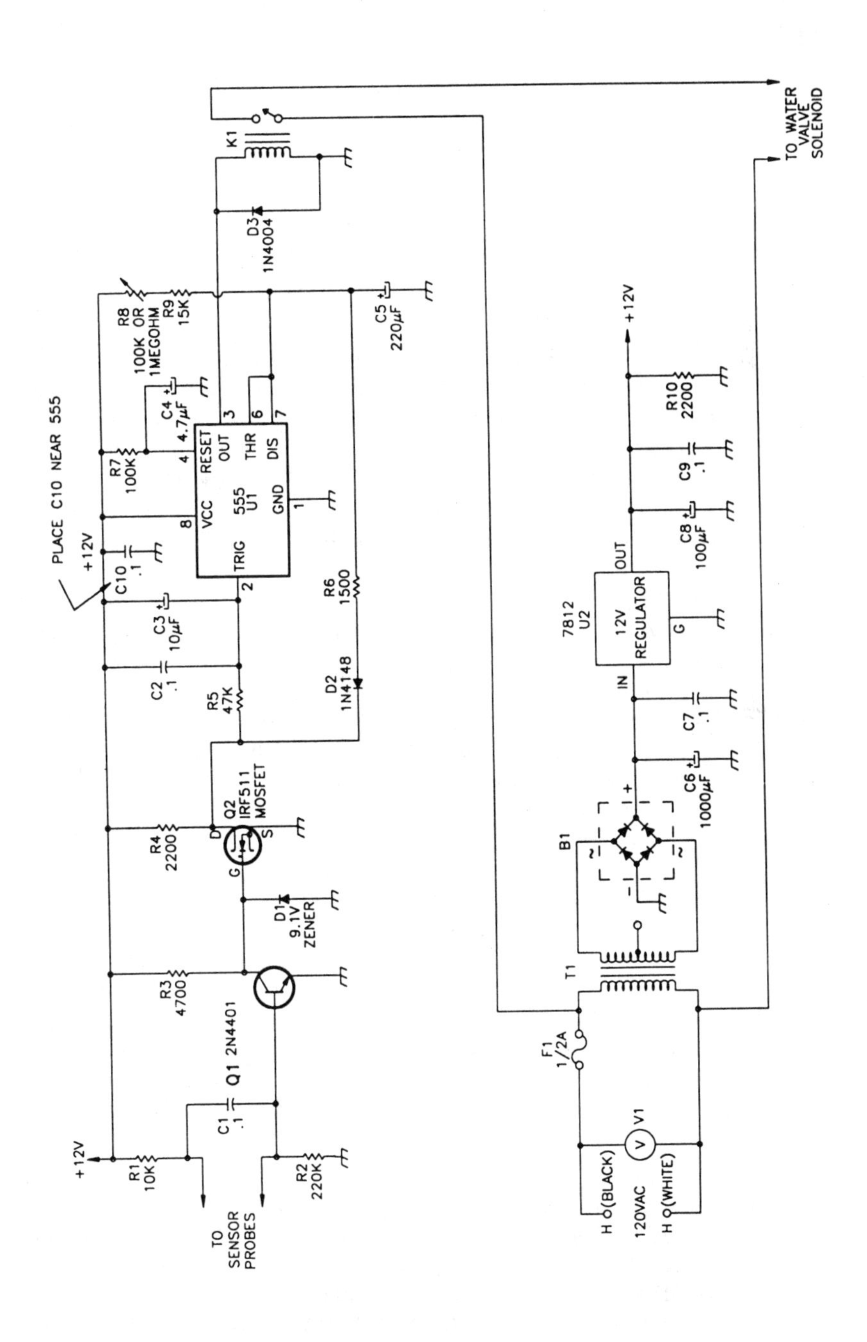

Figure 1 Complete schematic of the replenishment system.

The detector circuit consists of Q1, Q2, R1-R4, C1, and D1. Looking at figure 1, we see that when the probes are submerged in water, a small current will flow from +12 v through R1, through the water, and through the base of Q1. This current is amplified by transistor Q1. Although this current flow is small, the 2N4401 transistor has a minimum gain of 100, and with the 4.7K resistor on its collector, it should require a maximum base current of only 25 microamperes or so to drive it into saturation. Although the mineral content will cause the conductivity of water to vary, we found that our circuit was sensitive enough to function even in distilled water. As long as the probes remain submerged, the collector of Q1 will have only a fraction of a volt on it, held in saturation by the current flow through the water.

Q2, an IRF511 MOSFET, is inserted to provide a stage of inversion. The 555 is triggered by a signal going below one-third of the supply voltage, but the output of Q1 is low in its standby mode, going high when the 555 needs to be triggered, so it is necessary to invert the signal. We could have simply changed the detector circuit to reverse this, but we also wanted this extra stage to provide a quick discharge path for the 555's timing capacitor, and we need some additional gain to guarantee that this will be done in a timely manner.

We chose a MOSFET because of its high threshold voltage. A standard bipolar transistor could begin to conduct as soon as the collector voltage of Q1 reached about .6 v, but the MOSFET will not turn on until this voltage has reached 3-4 v. We felt that this provided a little extra insurance in case Q1 had trouble maintaining saturation in water of exceptional purity. A power MOSFET is not needed here--just about any N-channel enhancement- type MOSFET will do. We chose the IRF511 only because of its ready availability.

Capacitor C1 is placed across the lines going to the probes. It helps to steady the current flow through Q1 and prevent response to any transients, introduced either through movement in the water, or through noise pick-up on the lines between the sensor and circuit. Diode D1 is a 9.1 v zener diode. It is not essential to circuit operation, but serves to provide additional protection to the gate of the FET. Since the gate of the FET can only withstand 20 v or so, these components are easily damaged by any transients which may get through the power supply. Although the IRF511 includes internal protection, and this additional component is not absolutely necessary, we felt it was a good idea. This circuit will be operating continuously in most applications (including thunderstorms), and high reliability is important for this project.

When the water level drops below the sensor probes, the base current flowing through Q1 stops, and R2 guarantees that Q1's base is held at ground potential. This cuts off Q1, causing the voltage on its collector to go high. It will be clamped at about 9 v because of D1. When the gate of Q2 is pulled high, it turns on, and the voltage on its drain terminal is suddenly pulled close to ground. When this happens, two things occur. The first is that R5 starts to pull the trigger terminal of the 555 low. Because of C3, however, it will take about 1/2 second before the 555 triggers. This delay insures that the demand for water is genuine, and not just due to a momentary disturbance in the water. C2 helps prevent the possibility of the 555 being triggered by any electrical transients. When the timer does finally trigger, it turns on relay K1, which in turn energizes a water valve to turn on the water.

The drain of Q2 going low also performs a second function. In addition to triggering the timer, it also prevents its timing capacitor, C5, from charging. As long as the drain of Q2 is low, it will discharge C5 through R6 and D2; therefore, the time delay will not begin to time out until the water makes contact with the probes. Then the water will continue to run for the length of the delay set by R8 and C5. This approach guarantees that the water will always turn off at the same level, regardless of how low the water was when the circuit turned on.

When relay K1 is energized, it supplies 120 v to the coil of a water inlet valve. This will cause the valve to open, allowing water to flow. Once the sensor has made contact with the water and the 555 times out, relay K1 de-energizes, turning off the valve. Diode D3 suppresses any voltages generated by the relay turning off.

The power supply for this project is also included in figure 1. The line voltage is reduced to about 12.6 vac by transformer T1. Bridge B1 rectifies this, and along with C6, produces a DC voltage. The regulator outputs a steady +12 v for the circuit. Capacitor C8 helps provide better transient response for the regulator, and should be placed close to it on the circuit board. Capacitors C7 and C9 provide bypassing, while R10 helps guarantee that a minimum current flow through the regulator is maintained.

CONSTRUCTION DETAILS

Interior view of a completed water replenishment system project.

Although the control circuitry could be kept inside of the house and wires run outside to the sensor and water valve, in most applications the entire system will be outside; therefore, an enclosure must be used that can be sealed up reasonably well against the weather. The box we chose was a Radio Shack case (#270-627). We used a barrier strip as an interface between the circuit and the wires going to the sensor, and used silicone rubber to seal around the power wire cable and the wires going to the water valve. Since our use for this project was to supply water for a dog, we mounted the project case inside of the dog house, which protected it from the weather.

The sensor probe construction could be as simple as two wires placed so that they will be submerged at the proper water level; however, for reliable operation, we recommend something a little more substantial than this. Keep in mind that if the sensor somehow is removed from the water (it could happen if an animal were to get caught up in the sensor wires and pull), the water will then turn on and continue to run; therefore, it pays to spend a little extra time designing and placing the sensor.

We used a printed circuit board, etched two vertical strips, and then coated the copper with solder. This coating is especially important if the water has additives such as chlorine or flourine, which will more readily react with the copper. We then attached wires to the pads, and connected these to the terminals on the barrier strip. The sensor could be mounted as one piece; however, you will get better performance if you cut it in half vertically, so that each electrode is mounted on a separate board. In addition, one of the electrodes should be installed so that its mounting support is always above the water level. If you do not, the sensor cannot react immediately, because any water on the board will be bridging the contacts until it dries. It is not necessary to do any etching to create a sensor; you can simply get a piece of copper-clad board and cut out two sections, and coat them with solder. Attach leads and fasten it to the container holding the water, and you are set.

There are a couple of things to think about when you go to mount the sensor probes. Keep in mind that the time delay to turn off the water will start when water reaches the bottom of the highest sensor. The sensor plates do not necessarily have to be mounted at the same height; one could even be continuously submerged. What we recommend, however, is that each probe be mounted so that the final water level will cover each sensor about halfway. How far down the water will drop before it turns on will then be determined by how tall the sensor is, and the final water level will be determined by the setting of R8. The reason that we do not want either sensor to be more than half covered is because these electrodes will become contaminated over a period of time. By applying a voltage to two electrodes in a solution tainted by various salts and minerals, we are in essence electroplating one of the electrodes. Calcium and other mineral compounds will tend to deposit on the other electrode. Cleaning these off with a scouring pad every couple of months is a good idea. If this were neglected over a long period of time, the build-up might eventually restrict the current flow enough to prevent the circuit from being able to detect the presence of water, which could cause the water to turn on when none is required. Even worse, it would not be able to determine when to shut it off. This could result in a runaway condition which could literally turn the water on for hours. Placing the sensor so that half of it is out of the water provides a built-in safety feature. Since the portion of the sensor not normally in the water will not be contaminated, it will start the time delay as soon as the water rises above its normal level. This will stop the runaway condition within the period of the time delay.

The sensitivity of the sensor can be greatly enhanced by either increasing the size of the probes, or by moving them closer together. For best circuit performance, do not attempt to make the probes too small, or place them too far apart. If you have a special application for this project where the fluid being detected has less conductivity than tap water, you can also increase the circuit sensitivity by raising the value of R3.

Once the circuit detects the need for water, the 555 turns on relay K1 in response. This relay is used to turn on a solenoid- controlled water valve. The valve we used was a water inlet valve for a late model Sears washing machine (Sears part #388328 for a model #110.82881110 washing machine). A valve can also be obtained from an old washing machine that has been junked, but the price for the new one seemed too reasonable to justify taking a chance on a used part. We found the washing machine valve particularly convenient because of its garden hose compatible fittings and ready availability. These valves come with two inlets and one outlet. Each inlet has its own individually controlled 120 v solenoid. For this application, only one is necessary. You should be aware that the unused valve may dribble when the other valve is turned on, but this should not be a problem since this device will only be used outside. Make certain to carefully insulate the terminals of the water valve's solenoid, because there will be 120 v present when water is being added.

The outlet from this valve is not directly garden hose compatible, but a hose can be attached and clamped to run to remote locations. In our application, however, we simply mounted the valve directly on the watering container with the outlet pointing down. In order to maintain use of the outdoor faucet to which this was attached, we added a 'Y' adapter with individual cut-off valves. We used a section of garden hose and attached two female ends to interconnect the inlet valve to the 'Y' adapter on the faucet. If the inlet valve will be relatively close to the faucet, a washing machine hose can be used for this connection. Since we used this device to water a single dog, we only needed a moderately sized container. We trimmed a 5 gallon plastic bucket to meet our needs.

Resistor R8 is used to set the length of time the water will continue to run after contact is made with the sensor. It, therefore, sets the final water level. We used a 100K potentiometer for R8, which gave us an adjustment range of about 4-25 seconds. For larger water containers, which might be used for livestock, a longer delay may be desirable. Using a 1 meg potentiometer for R8 will allow a run time of around 4 minutes.

Although this circuit has been designed for maximum reliability, and our prototype has operated flawlessly, a failure is always possible. When designing any circuit, it is important not only to consider how the circuit will respond when it is operating correctly, but also how it will respond if it fails. In the case of this particular device, the most probable failure mode would be for the valve to not turn on when required; however, there is a remote chance of a failure in which the valve would stay on. If there is any possibility of damage as a result of this, you may want to add an additional "fail-safe" circuit to guarantee water shut-off (figure 2 shows one possible circuit). The normally closed contacts of the relay in this circuit could be placed in series with the water valve solenoid, and the sensor probes placed above the main probes. We did not feel the need to install this on our system, however, since the probability of this type of failure is extremely small.

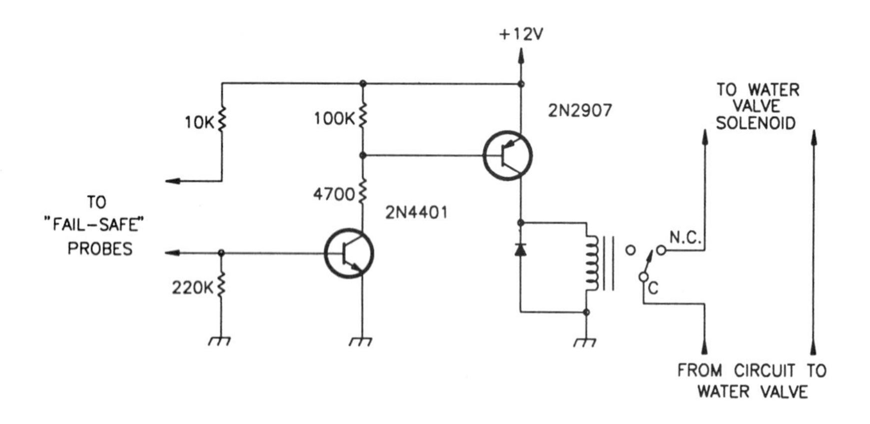

Figure 2 An optional fail-safe circuit.

If you live in a climate where the temperature may go below 32°F while this device will be in use, don't forget to take steps to prevent pipe and hose freezing.

PARTS LIST

Semiconductors:

B1- 100 v-1.4 amp bridge rectifier (Radio Shack #276-1152)
D1- 1N4739-9.1 v zener diode-1 w (Radio Shack #276-562)
D2- 1N4148 diode (Radio Shack #276-1122)
D3- 1N4004 silicon rectifier (Radio Shack #276-1103)
Q1- 2N4401 NPN silicon transistor (Radio Shack #276-2058)
Q2- IRF511 MOSFET (Radio Shack #276-2072)
U1- 555 Timer IC (Radio Shack #276-1723)
U2- 7812 +12 v regulator IC (Radio Shack #276-1771)

Capacitors:

C1, C2, C7, C9, C10- .1 ufd-50 v disc
C3- 10 ufd-16 v electrolytic
C4- 4.7 ufd-16 v electrolytic
C5- 220-ufd-16 v electrolytic
C6- 1000 ufd-25 v electrolytic
C8- 100 ufd-16 v electrolytic

Resistors: (All resistors 1/4 w unless stated otherwise)

R1- 10k
R2- 220k
R3- 4.7k
R4, R10- 2.2k
R5- 47k
R6- 1.5k
R7- 100k
R8- 100k or 1 meg potentiometer (see text- we used Radio Shack #271-284)
R9- 15k

Miscellaneous:

F1- 1/2 amp fuse and holder
K1- 12 v relay (Radio Shack #275-248)
T1- 12.6 v-450 mA power transformer (Radio Shack #273-1365)
V1- varistor (Radio Shack #276-570)
Circuit board (we used a Radio Shack #276-168)
Project case (we used a Radio Shack #270-627)
2-terminal barrier strip (Radio Shack #274-656)
Water inlet valve (see text- we used a Sears part #388328 from a Sears washing machine model #110.82881110)
Line cord and misc. hardware

BRINGING UP THE UNIT

Once the circuit has been put together, measure the resistance from the prongs of the line cord to DC circuit ground. There should be a reading of infinite resistance here. If not, locate the problem before proceeding. It is very important that this circuit be isolated from the power line, since the probes will be submerged in water that animals, and probably humans, will be coming in contact with.

Once the safety of the unit has been verified, connect a 22K or 33K resistor across the input terminals for the probe, and then plug the circuit in. The relay should remain de-energized. With R8 set to its position of minimum resistance, disconnect one lead of the 22K or 33K resistor from its input terminal. After about a 1/2 second delay, the relay should pull in. Reconnect the resistor lead, and the relay should drop out in about 3-

4 seconds. If the device seems to be working correctly, go ahead and install it in its permanent location, and attach the sensor leads and water inlet valve. Apply power once again, with the water turned on. The water should turn on very shortly after power is applied. After the water has risen to the level necessary to contact the sensor probes, it should continue to run for the additional 4 second delay time and cut off. You can then adjust the potentiometer by alternately turning the adjustment and dipping water out, forcing it to run another cycle, until the water level stops at the desired location. The final water level can also be adjusted to some degree by restricting the flow of water at the faucet. There is no reason that the water must flow wide open.

If the unit does not seem to function properly, check the following:

- Before checking anything else, verify that the power supply is operating correctly. You should read very close to 12 v across pins 8(+) and 1(-) of the 555 IC. If not, measure the voltages at the input and output of the regulator IC. If the voltage on the input is zero, check the wiring of the bridge and transformer, and make certain that fuse F1 has not blown. If there is at least some voltage on the input, but the output voltage is low, make certain that Vcc and ground have not been shorted by a solder bridge. Also make sure that you have not used the center tap lead of the transformer secondary by mistake. If the voltage going into the regulator is correct, but the voltage coming out is low, and no short is detected across the power supply rails, replace the regulator.
- If the power supply voltages are correct, but the relay does not pull in or release when it should, then take a voltage measurement from the collector of Q1 to ground with the probe terminals both open and shorted. With the probe terminals open, the voltage on Q1's collector should be around 9 v. With the probe terminals shorted, the voltage should read less than 1/2 v. If either of these readings are off, check the wiring around Q1 very carefully for any errors. Make certain that D1 has not been installed backwards, since this would prevent the voltage on Q1's collector from rising above .6 v. If the voltage stays either high or low, and no errors can be found, replace Q1. If the voltage stays low and will not rise, check to make sure that neither D1 or Q2's gate is shorted to ground.
- If the voltage on Q1's collector switches correctly, but the relay does not respond, measure the voltage on the drain terminal of Q2, and, once again, open and short the probe terminal connections. When the probe terminals are shorted, the drain voltage should be very near +12 v, going close to ground when the terminals are open. If not, make certain that all wiring around Q2 is correct, and that the gate terminal of the FET is in fact connected to, and following, the voltage on Q1's collector. And make sure that D2 is not installed backwards. If Q2's drain does not switch and no other cause can be found, replace Q2.

- If the voltage switches correctly on the drain of Q2, next measure the voltage on pin 2 of the 555 with the probe terminals open. This voltage should drop close to zero, going high again when the probe terminals are shorted. If it does not, check the network of R4, R5, and C2 very carefully for errors. If it appears that pin 2 of the IC is internally shorted high or low, you may have to replace the IC. If pin 2 does follow the signal correctly, but the relay is not energizing at all, make certain that the voltage on the reset terminal (pin 4) is about +12 v. If the relay pulls in but never releases, open the probe terminals until the relay pulls in, and then short the terminals. Then measure the voltage on pin 6 or 7 of the 555. The voltage should rise to about 8 v, at which point the timer should reset. If the voltage never reaches this point, check the connections on R8 and R9. Make certain that the potentiometer is not open. If these connections are good, you either have a timing capacitor (C5) with excessive leakage or a 555 that is bad. If the voltage reaches 8 v, but the circuit does not reset, first make certain that pin 2 is in fact high, as it should be, and then replace the 555.
- If the relay does pull in and drop out, but the water level does not always stop close to the same point, check D2 and R6 for bad connections. If D2 is installed backwards, the water level will stop rising the second it makes contact with the sensor, rather than continuing on for the correct delay period.
- If the timer ever shows a tendency to turn back on for a second cycle immediately after turning the valve off, the 555 is probably being triggered by electrical noise generated by the relay contacts arcing on cut-off. The easiest solution is to kill the noise at its source. Generally, soldering a varistor similar to V1 across the relay contact terminals will solve problems of this nature. Additional .1 ufd capacitors across the supply lines may also help; however, with this circuit, and the type of water valve that we used, it is not likely that you will encounter any problems of this sort.

DEMAND-OPERATED PLANT WATERING SYSTEM

This system is designed to automatically supply water to your lawn, garden, or flower bed. The circuit is connected to probes placed beneath the ground, which are used to detect the moisture content of the soil. This makes the device aware of when additional water is needed. Although the system can be made to supply water immediately, it is generally considered preferable to water in the morning; therefore, we have included a photo-sensitive detector. When the circuit senses that more water is needed, it waits until the following dawn before it turns it on. When the sensor probes indicate that sufficient water has been delivered, the circuit shuts the water off.

This system is obviously superior to a plant watering system based solely on a timer, which is incapable of sensing whether or not additional water is actually needed. This prevents the possibility of damaging the plants due to wasteful overwatering. It also deprives your neighbors of the laugh they would get watching your timed sprinkler system operating in the rain.

CIRCUIT DESCRIPTION

Although this project performs a function for plants quite similar to what the 'Water Replenishment System' does for animals or pools of water, a glance at figure 1 will show that this circuit is totally different. Both circuits are required to turn on water when needed, but that is really about all that they have in common. When the requirements for the two projects are closely examined, it turns out that they have many more differences than similarities.

The most important difference is in the method that we must use to determine when additional water is needed. In the other project, all we needed to detect was whether water was present or absent. It was either there or it wasn't. In this system, the situation is not so clear-cut. There will always be some moisture in the soil. Our circuit must be able to detect varying degrees of moisture, and be capable of determining when a threshold level has been exceeded.

The schematic in figure 1 shows how this was accomplished. The moisture detector is composed of the 741 op-amp, C1, and resistors R1, R2, R3, R4, and R5. Two sensor probes (the construction of which are discussed in the 'Construction Details' section) are buried in the ground that is to be watered. The resistance across these two probes will tend to vary with the level of moisture in the soil. The higher the content of water, the lower the resistance across the probes will be. The probe inputs, therefore, act as a variable resistor whose resistance decreases with more

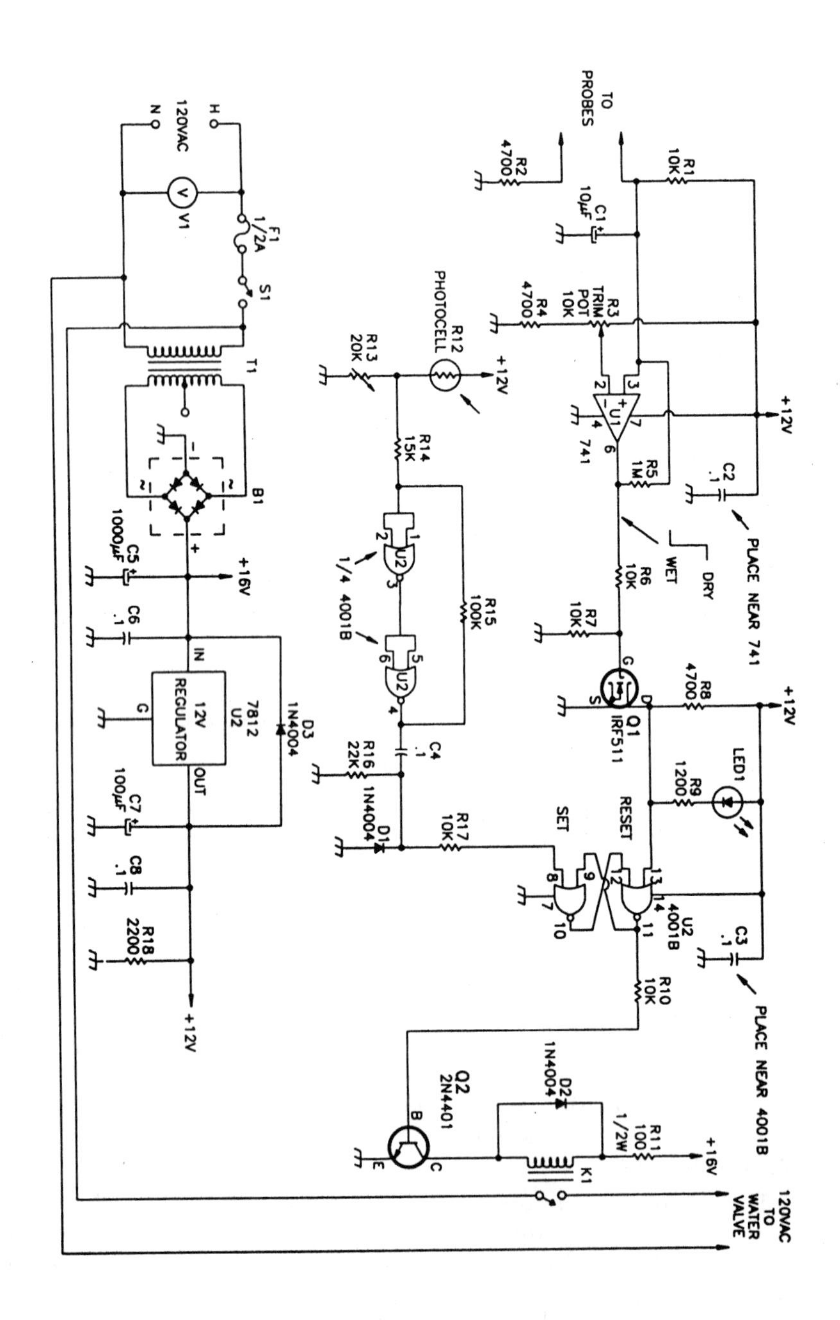

Figure 1 Schematic of the plant watering system.

water, and increases with less water. By placing the probe input connections in series with R1 and R2, we form a voltage divider, which connects to pin 3 (the non-inverting input) of the 741 op-amp. As the ground becomes drier, the voltage on pin 3 will rise. As water is added, the resistance will drop across the probes, and the voltage on pin 3 will drop. C1 helps to stabilize this voltage, and also holds pin 3 momentarily low on power-up. This will cause the circuit to always power-up with the water valve turned off.

The 741 is connected to run as a comparator. There is no negative feedback provided, so it is running at full gain. As a result, pin 6 of the op-amp will almost always be at its maximum high or low output voltage. For a bipolar op-amp like the 741, the low output will be about 1-2 v, and the maximum high output on a 12 v supply will be around 10-11 v. Whether pin 6 is high or low is determined by the relationship between pins 2 and 3. When pin 3 has a higher potential on it than pin 2, the output will be high. When pin 2 has a higher potential than pin 3, the output will be low. In order to make the voltage on pin 2 variable, we connect it to a potentiometer, R3, which is placed across the supply voltage in series with R4. We then adjust R3 so that the voltage on pin 2 of the 741 will match the voltage on pin 3 when the ground is just approaching the point where water will need to be added. By doing this, the output of the 741 will be low when there is sufficient moisture in the soil, swinging high when more water is needed.

The purpose of resistor R5 is to provide a limited amount of positive feedback. This insures that the output will not hesitate when the voltages on pins 2 and 3 are very close together. As long as the voltage on pin 3 is below that on pin 2, the output of the 741 will be low, but as the pin 3 voltage slowly rises, a point will be reached where the output will start to go high. When these two voltage are very close, the state of the output is unpredictable. With the positive feedback provided by R5, however, the first time the output starts to go high, the voltage on pin 3 is pulled up an additional amount, eliminating the possibility of hesitation on the output of the 741 (for a more detailed description of how this works, refer to the 'Circuit Description' of the 'Electronic Thermostat' project).

One effect produced by adding the positive feedback is a differential between the wet and dry switching points. The probe resistance necessary to make the output switch from low to high will not be the same as when it switches the other way. This allows a means of obtaining independent adjustment between the cut-on and cut-off points. By substituting a 1 meg potentiometer in series with a 100K resistor in place of R5, this differential can be made variable. R3 is used to set the point at which the circuit recognizes a need for water, and R5 sets the point at which the demand for water is satisfied. This allows for a variable run time without changing the depth of the probes. We tried this in our circuit, and found that if the probes are at a relatively shallow depth, as ours were, the difference was too insignificant to justify including the potentiometer. The soil resistance would change so rapidly when water was added that the difference in switching points would have little effect; however, if your probes will be placed deeper for a longer watering time, such as for fruit trees, the water will tend to trickle down to the sensor level at a slower rate. In this case, the use of a potentiometer for R5 may be justified.

When the output of the 741 is low, the voltage present on pin 6 will be less than 2 v. The voltage divider of R6 and R7 cuts this in half, so that the voltage applied to the gate of Q1 will be no greater than 1 v. Since this FET requires a gate voltage of 3 v minimum to turn on, this guarantees that Q1 stays off. When the output of the 741 switches high, this applies a minimum of 5 v to the gate of the FET, turning it on. This turns on LED1, which serves as an indicator that water is needed. It also enables the flip-flop, composed of the two NOR gates, to turn the water on the following morning. R8 guarantees that the voltage on pin 13 of U2 will be at the full 12 v DC potential when the FET is off. Without R8, any leakage current through the FET might cause a voltage drop across LED1, reducing the input voltage to the flip-flop.

If all we were after was to turn the water on as soon as a need was detected, we could simply insert a relay in place of R8; however, since it is generally preferable to water early in the morning, we need some means of triggering the circuit at this time. The method that we chose to accomplish this was to use a photosensitive circuit. The photocell, R12, forms a voltage divider with potentiometer R13. As the light intensity increases, the voltage across R13 will rise; however, the voltage across the resistor is a slowly changing analog value. Just as with the moisture probes, we need to convert this to a digital signal. Although another op-amp could be used to perform this function, we chose to use the remaining gates in the 4001 IC, since they were already available. We combined these two gates with R14 and R15 to form a simple "Schmidt" trigger circuit.

To understand how this circuit works, let's start out by assuming that the photocell is dark. The voltage across R13 will be relatively low in this case. As an example, when this voltage is at 2 v, the voltage on pins 1 and 2 of the 4001 IC will also be close to this value. Since this is well below the halfway point at which CMOS gates normally switch, pin 3 will be high, because the gate functions as an inverter. The second gate inverts the signal again, back to its original state, and adds additional gain; thus, in this case, pin 4 of the 4001 will be near ground potential. This means that the end of R15 connected to pin 4 is also at ground; therefore, the voltage on pin 1 will be less than the voltage across R13, due to the voltage divider set up by R14 and R15. In order for pin 1 to reach the 6 v level necessary to switch the output, the voltage across R13 must, therefore, exceed this level, with the precise amount determined by the relative values of R14 and R15.

As the light intensity increases, the voltage across R13 rises until the voltage on pin 1 reaches the switching point (about 1/2 Vcc). When this happens, pin 4 goes high, reversing the polarity of the voltage divider. The voltage on pin 1 will now be higher than that across R13. As a result, the voltage will have to drop considerably across R13 before the circuit can switch back the other way. This gives us a circuit with a digital output, and sufficient hysteresis to prevent false triggering due to shadows or clouds passing over.

The output of the photoelectric circuit will switch low at dusk and high at dawn. To make an operational system, we need a circuit that will trigger on a low to high transition. We started by building a flip-flop out of the two remaining NOR gates. The output of the photo circuit is coupled to the 'set' terminal of the flip-flop through a capacitor, C4. This capacitor couples a momentary positive pulse to pin 8 of U2 at the time the output switches from low to high. If the ground already has sufficient water, pin 13 will be high, and pin 11 will remain low at the time that the pulse is applied to pin 8. If the soil is in need of water at the time that the pulse is applied to the 'set' terminal, pin 13 ('reset') will be low, and the flip-flop will reverse states, with pin 11 going high. This will turn on transistor Q2, which subsequently turns on relay K1 and the water valve. The circuit will remain latched on until the ground has sufficient water, causing the 'reset' terminal, pin 13, to go high. This will cause the flip-flop to reset, taking pin 11 low, and turning off relay K1 and the water valve.

The purpose of D1 and R17 is to protect the input on pin 8 of the 4001 IC at dusk when the photo circuit switches from high to low. Since R16 is already holding pin 8 at ground potential, applying a negative pulse through C4 will take pin 8 several volts below ground, possibly damaging the input. Diode D1 clips the negative signal so it cannot go more than .6 v below ground. R17 limits the current so that the IC's internal protection diodes can safely handle the reverse voltage, as well. Diode D2 suppresses any inductive voltage generated when relay K1 de-energizes. We also inserted R11 in series with the relay coil, in order to reduce the 16 v to about 12 v.

Switch S1 is an optional on/off switch. 120 v is applied to the circuit through fuse F1, transformer T1 reduces the AC voltage to about 12.6 v, and rectifier bridge B1 converts this AC waveform to a pulsating DC, which is filtered by capacitor C5. C6 provides bypassing. The 7812 regulator IC maintains a steady output of +12 v. Capacitor C7 provides for better transient response, while C8, C2, and C3 all provide bypassing. C2 and C3 should be placed as close as possible to the 741 and 4001 ICs, respectively. R18 provides a minimum current draw through the regulator, while diode D3 protects the regulator in the event that the input voltage drops faster than the output voltage on power-down.

If you tend to prefer rather long watering times for your plants, you may wish to consider modifying this circuit by adding a re-triggerable timer to the output stage. This would cause the system to add water until the sensor probes indicated the need had been met, and then it would allow the water to run for an additional period determined by the time delay. This not only would allow more water to be added, but it would allow the final watering level to be adjusted by a potentiometer, rather than having to move the probes. One simple way to do this is to use the timer stage from the last project (the 'Water Replenishment System'). The 555 timer stage in that project (along with D2 and R6) can be interfaced to this device by connecting the 555 input (and D2) to the collector of Q2 in this project. The network of K1, D2, and R11 can be replaced with a 4.7K resistor going to +12 v. Using a 1 meg potentiometer and a 1000 ufd capacitor for the timing elements should allow an additional run time of nearly 20 minutes.

CONSTRUCTION DETAILS

Interior view of a completed plant watering system project.

As with the water replenishment system in the previous chapter, the control circuitry for this project could be placed either inside or outside of the house. If it will be placed outside, a weatherproof enclosure must be used. We used a Radio Shack plastic case (#270-224). We found that a barrier strip made an excellent interface between the circuitry and the wires going to the sensor. If the circuit itself will be inside the house, the photocell will have to be mounted in a window, facing outside, so that it can sense the exterior light level. The barrier strip can then also be used to provide the circuit connections for the photocell. Our system was completely installed outside, so we mounted the photocell directly on the box itself. If you do place the project case outside, mount the barrier strip on the bottom, to reduce the effects of weather on the terminals. Bring any other wires in through the bottom of the box, and seal around them with silicone rubber.

The sensor probes for this project were two 1/2" copper pipes, about 12" long. They were buried about a foot apart, with the wires attached using ground clamps. For our application, the sensor probes were buried 2-3" deep; however, depth and sensor placement will be largely dictated by the type of soil and the length of watering time desired. Generally, the deeper the probes, the longer it will take the water to reach them, and hence, the longer the watering. As a general rule, a good starting point is to place the probes just below root level of the plants you are watering. Make certain to construct the two probes out of identical materials. If you make one probe out of a length of copper, and the other from another type of metal, the dissimilar metals may react with the acid in the soil to create a battery.

The wires connecting the sensor probes to the circuit input on the barrier strip should be placed at least a couple of inches underground. This will prevent the probes from being disturbed, and keep the wires from being damaged.

To supply the water for our prototype, we used a 'Y' coupling at an outside spigot, and connected a length of hose with a female coupler at each end. If the water valve will be very close to the spigot, a washing machine hose can be used. If not, then a standard garden hose can be modified by cutting it to the proper length and adding a female end.

When relay K1 energizes, it supplies 120 v to a valve, which turns on the water. Although there are many types of electrically operated water valves available, we used a water inlet valve from a Sears washing machine (model #110.82881110 part number 388328). We found the washing machine valve particularly convenient because of its garden hose compatible fittings, ready availability, and low cost. These valves come with two inlets and one outlet. Each inlet has its own individually controlled 120 v solenoid. For this application, only one is necessary. You should be aware that the unused valve may leak a small amount when the other valve is turned on, but this should not be a problem since this device will only be used outside. Make certain to carefully insulate the terminals of the water valve solenoid, since there will be 120 v present when water is being added.

The outlet from this valve is not directly garden hose compatible, but a hose can be attached and clamped in order to run it to remote locations. Your hose could supply water to a sprinkler, or any other type of watering device for the lawn. In our case, we routed the hose through the flower garden and plugged the end, and then drilled 1/8" holes through the hose where we wanted the water. Although this valve sufficed for our application, it has a restricted flow that may not reliably operate a heavy duty sprinkler. For heavier loads, a valve from a sprinkler supply company may be required. If necessary, several valves can be driven by K1 for multiple sprinklers.

Since this project is a closed loop system, it is essential that the probes be able to properly sense the presence of the water as it is being added. Just as the circuit turns on the water by detecting a high soil resistance, it must detect a lower soil resistance to turn the water off, therefore, the device that applies the water must do so in such a way that it insures that the sensor probes are receiving a relatively typical amount of water. Keep in mind that if the location of the sprinkler is moved, either accidently or on purpose, in such a way that water does not reach the sensor, the water will not shut off. The integrity of the wiring and connections between the circuit and sensor probes are of equal or greater importance. If the sensor lines become open, the circuit will respond by not sensing that water has been added. For proper operation, make certain that all connections are secure and that the wires connecting to the probes are adequately protected.

PARTS LIST

Semiconductors:

B1- 100 v-1.4 amp bridge rectifier (Radio Shack #276-1152)
D1, D2, D3- 1N4004 silicon rectifier-400 v-1 amp (Radio Shack #276-1103)
Q1- IRF511 MOSFET (Radio Shack #276-2072)
Q2- 2N4401 NPN silicon transistor (Radio Shack #276-2058)
U1- 741 op-amp (Radio Shack #276-007)
U2- 4001 quad NOR gate (Radio Shack #276-2401)
U3- 7812-12 v regulator (Radio Shack #276-1771)

Capacitors:

C1- 10 ufd-16 v tantalum
C2, C3, C4, C6, C8- .1 ufd-50 v disc
C5- 1000 ufd-25 v electrolytic
C7- 100 ufd-16 v electrolytic

Resistors: (All resistors 1/4 w unless stated otherwise)

R1, R6, R7, R10, R17- 10K
R2, R4, R8- 4.7K
R3- 10K trimmer potentiometer (Radio Shack #271-343)
R5- 1 meg fixed resistor or 1 meg potentiometer in series with 100K fixed resistor (see text)
R9- 1.2K
R11- 100 ohm-1/2 w
R12- cadmium sulfide photocell (Radio Shack #276-118)
R13- 20K trimmer potentiometer (Radio Shack #271-340)
R14- 15K
R15- 100K
R16- 22K
R18- 2.2K

Miscellaneous:

F1- 1/2 amp fuse and fuseholder
K1- 12 v relay (Radio Shack #275-248)
S1- 120 v toggle switch **Optional
T1- 12.6 v-450 mA transformer (Radio Shack #273-1365)
V1- varistor (Radio Shack #276-570)
2-position barrier strip
Circuit board (we used a Radio Shack #276-168A)
Project case (we used a Radio Shack #270-224)
Electrically-operated water valve (we used a water inlet valve from a Sears washing machine-part number 388328)
Components for sensor probes and water distribution, line cord, circuit board standoffs, misc. hardware

BRINGING UP THE UNIT

Before operating the system, a check for power line isolation from DC circuit ground should be performed. This is important not only for safety, but also to prevent disruption of the sensor circuit. An ohmmeter reading from DC circuit ground to either prong of the line cord should reveal infinite resistance.

When installing the sensor probes, the depth and the soil type will be major factors in circuit operation. Some experimentation in probe placement may be necessary for optimum performance. As a general rule, the deeper the probes, the longer the water will run. For light, frequent watering, the probes should be relatively close to the surface, maybe 3-4" deep. For a longer watering, the probes should be farther down. After the probes have been installed, consistent readings cannot be obtained until the soil has been tightly packed around them. The surest way to do this is to soak the disturbed area with water. You may want to do this two or three days before you install the circuit, since it cannot be calibrated until the ground has dried.

Prior to using this circuit, the potentiometers must be adjusted, but since the conditions will vary from one installation to another, the exact settings will have to be determined experimentally. If you used a 1 meg potentiometer for R5, start by setting it to its position of maximum resistance. The 20K potentiometer used for adjusting the light sensitivity should be set so that pin 4 of the 4001 IC reliably switches high in the morning and low at dusk. There should be no tendency for the circuit to trigger due to shadows or clouds passing over; we found a setting of about 8K to be a good choice. You may want to set the potentiometer to this value before installing it on the board. As a starting point for R3, which determines how dry the soil must be before the water will be turned on, we found that a reading of 5.0 v on pin 2 of the 741 op-amp was about optimum; however, the final setting for this potentiometer could vary considerably from one installation to another, depending upon the characteristics of the soil and the exact construction and placement of the sensor.

After powering up the unit, you should wait a couple of hours before attempting to calibrate the system. This will give the voltage across the sensor probes time to stabilize. Wait until the first day that you are sure that watering will need to take place on the following morning. At that time, check to see that LED1 is lit. If it is not, rotate potentiometer R3 (in the direction that decreases the voltage on pin 2 of the 741) until the LED comes on, and then continue for another one-fourth to one-half turn. This LED, when lit, indicates that the probes detect a need for water. Assuming that no water is added through rainfall or other means during the night, the system will turn on in the morning. After observing the operation for a few days, you can adjust R3 to your specific watering needs. Turning R3 to decrease the voltage on pin 2 of the 741 will cause the system to turn on more often; going the other way will decrease the frequency of watering.

If you used a potentiometer for R5, you can decrease its resistance for a longer watering time, if necessary. The effect produced by R5 on the circuit will be greater at the lower end of its range; therefore, to see a significant change, start by turning the potentiometer at least halfway towards its minimal position. Turning beyond that halfway point will cause the response to be more dramatic; however, its effect will be minimal if the probes are not buried very deep. The depth of the probes will have a greater effect on the watering time than the adjustment of R5. Also, adjusting R5 may have an effect on the adjustment of R3, as well.

If the unit does not seem to function properly, check the following:

- Check the power supply voltages first. There should be over 15 v on the input of the 7812 regulator IC, and very close to 12 v on the output. If there is no voltage on the input, check fuse F1 and the wiring of the transformer and bridge. If the voltage is low on the output, check to make sure that the power supply lines have not been shorted through a solder bridge. If the output voltage is low, but the supply lines are not shorted, and the input voltage is normal, replace the regulator. Make certain that the 12 v supply voltage is not only correct, but is, in fact, reaching the supply pins of both the 741 and 4001 ICs.
- If the power supply voltages seem correct, disconnect the wires going to the sensor probes, and short the terminals on the barrier strip. LED1 should be off. Remove the short, and LED1 should come on. If this portion is working correctly, go to step 3 below. If the LED stays on or off, then there is a problem in the moisture sensing circuit. Start by measuring between pins 2 and 4 of the 741. This voltage should be in the range of 3.8-12 v. As R3 is rotated, the voltage should go from one extreme to the other. The voltage between pins 3 and 4 should be around 3.8 v with the sensor inputs shorted, and near 12 v with them open. With R3 set about mid-range, measure the voltage from pin 6 of the 741 to ground with the sensor inputs open. The voltage here should be around 10-11 v. Shorting the inputs should cause pin 6 to drop to under 2 v. If the input voltages are correct, but the op-amp doesn't switch, replace the 741. If the 741 output is switching correctly, check the voltage on the gate of the FET. With the input terminals shorted, there should be no more than a volt or so at this point. With the inputs open, there should be about 5 v on the gate. If these voltages are correct, but the LED still doesn't switch, measure the voltage on the drain of Q1. There should be about 12 v here with the input terminals shorted, and less than a volt with the inputs open. If the drain voltage does not change, make sure that the source terminal is properly grounded, and, if so, replace the FET.
- With the sensor input terminals open so that LED1 is on, momentarily short pin 8 of the 4001 IC to +12 v. Relay K1 should energize and stay on. Shorting the input terminals should cause K1 to de-energize. If this portion of the circuit also seems to work correctly, proceed to step

4 below. If relay K1 either will not come on or go off, then you have a problem in the flip-flop or relay driver stage. Check all wiring carefully on pins 7 through 14 on the 4001 IC for any errors. With LED1 on, make certain that the voltage on pin 13 is near ground potential. Short pin 8 of the 4001 to +12 v. The voltage on pin 11 should go high. If it does, but the relay does not turn on, the problem resides with Q2, the relay, or the associated wiring. Make sure D2 is good, and not installed backwards. By removing the short from pin 8 to +12 v, the voltage on pin 8 should drop to 0 v. Short the sensor input terminals so that LED1 goes off. This should cause pin 11 to go low. If the 4001 does not seem to respond as it should, and all wiring is correct, replace the IC.

- If everything else is working correctly, the photocell circuit may not be properly triggering the flip-flop. While measuring from pin 4 of the 4001 IC to ground, cover the photocell completely with your hand. The voltage on pin 4 should be very close to 0 v. Removing your hand should cause the voltage to rise to almost 12 v, assuming there is sufficient light, and that R13 is properly adjusted. If there is no change in the output voltage, measure the voltage on pin 1 or 2 of the 4001 with respect to ground. This voltage should go up and down as you cover and uncover the photocell. If it does not, check the photocell and R13 to make sure they are good. If the voltage does go up and down with the light intensity, but the output on pin 4 doesn't change, check the wiring of the 4001, R14, and R15. If all seems to be correct, replace the IC. If switching is occurring at pin 4, but it does not trigger the flip-flop, check the connections between pins 4 and 8. Make sure diode D1 is not installed backwards. If everything seems to be correct, try replacing C4.

TELEPHONE REMOTE CONTROL AND CALL SCREENING DEVICE

This is a dual-purpose device that can be configured to either function as a telephone call screener, or as a telephone remote control system. When built to serve as a call screener, this device (when enabled) will answer any telephone call on the first ring. It will then allow the caller about 10 seconds to press a key, which will identify them as someone to whom you have given the code. If they press the correct key in time, an alarm will sound to indicate this, and the line will be held for some additional time so that the phone can be answered. If a key is not pressed, or the wrong one is, the device will then hang up in 10 seconds. This will keep you from being unnecessarily bothered by anyone that you don't want to talk to.

For our application, we found that there were many instances when we did not want to be disturbed. Every time the phone rang, however, we pictured our wives standing on the side of the road at a payphone with a disabled car and two kids, but every time that we did answer the phone, it was a salesman trying to sell us windows, or offering us our 37th 'free' trip to Hawaii. They were wasting our time. This project will allow you to screen the calls without the hassles involved with an answering machine (having to listen to your own recording, and then to the response of the caller). This device would be especially useful for anyone receiving harassing phone calls.

With a slight modification to the circuit, this project can also be used to turn an electrical device in your home on or off. When the unit answers, and the correct key is pressed, a connected device can be controlled from a remote location. For example, you may want to turn on your air conditioner before you leave work, or to turn on a coffee maker with your car phone a few minutes before you arrive home. Another application might include controlling devices in locations that are often unmanned. Examples might include turning the heat up in a vacation home, or in a church building prior to a meeting.

CIRCUIT DESCRIPTION

In order for our circuit to perform what is expected of it, it must be able to detect when the phone is ringing, answer the call, and hold the line for a specific period of time while listening for the control tone. If the tone is received, it must set off an alarm and hold the line for some additional time, giving the person a chance to get to the phone. If the device will be used for remote control, then the alarm is not needed, but some type of latch or timer circuit will still be needed for the device that is to be controlled.

Figure 1 shows a block diagram of what we will need to accomplish this. It is often convenient when designing more complicated circuits to first break the overall project down into smaller stages, or tasks, as shown here, and then design the circuit for each block. Naturally, care must be taken with this approach to make certain that each stage will properly interface with the one both preceeding and following it. In our block diagram, we start with a ring detector circuit. The function of this circuit is to detect the presence of a ring signal coming over the phone line. The output of this stage is normally high, going low when a ring is present. The next stage, a 10-12 second delay, will energize a relay, K1, to answer the call. This is done by connecting a telephone interface transformer to the telephone line. After the delay has expired, the timer will de-energize K1, disconnecting the transformer from the phone line; however, if a valid control tone was received over the line during the time that K1 was energized, the tone decoder will have already set off the alarm and triggered a second timer to hold the phone line an additional 50 seconds. In remote control applications, this additional hold time will not be required, and this stage will be omitted. In addition, the alarm stage will be replaced with either a timer or latch circuit to turn the controlled device on or off. With either configuration, of course, we will also need a power supply.

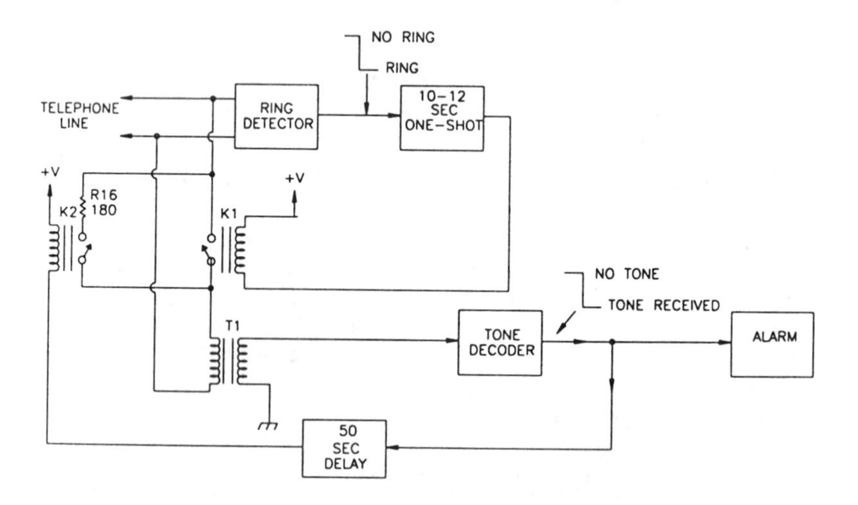

Figure 1 A block diagram of the system.

Now that we have outlined what must be done, let's take a look at figure 2. This is the schematic for the prototype which we built, which functions as a selective answering device. Although it is a little more complicated than many of the other projects in this book, it is really quite easy to understand if you examine it one stage at a time.

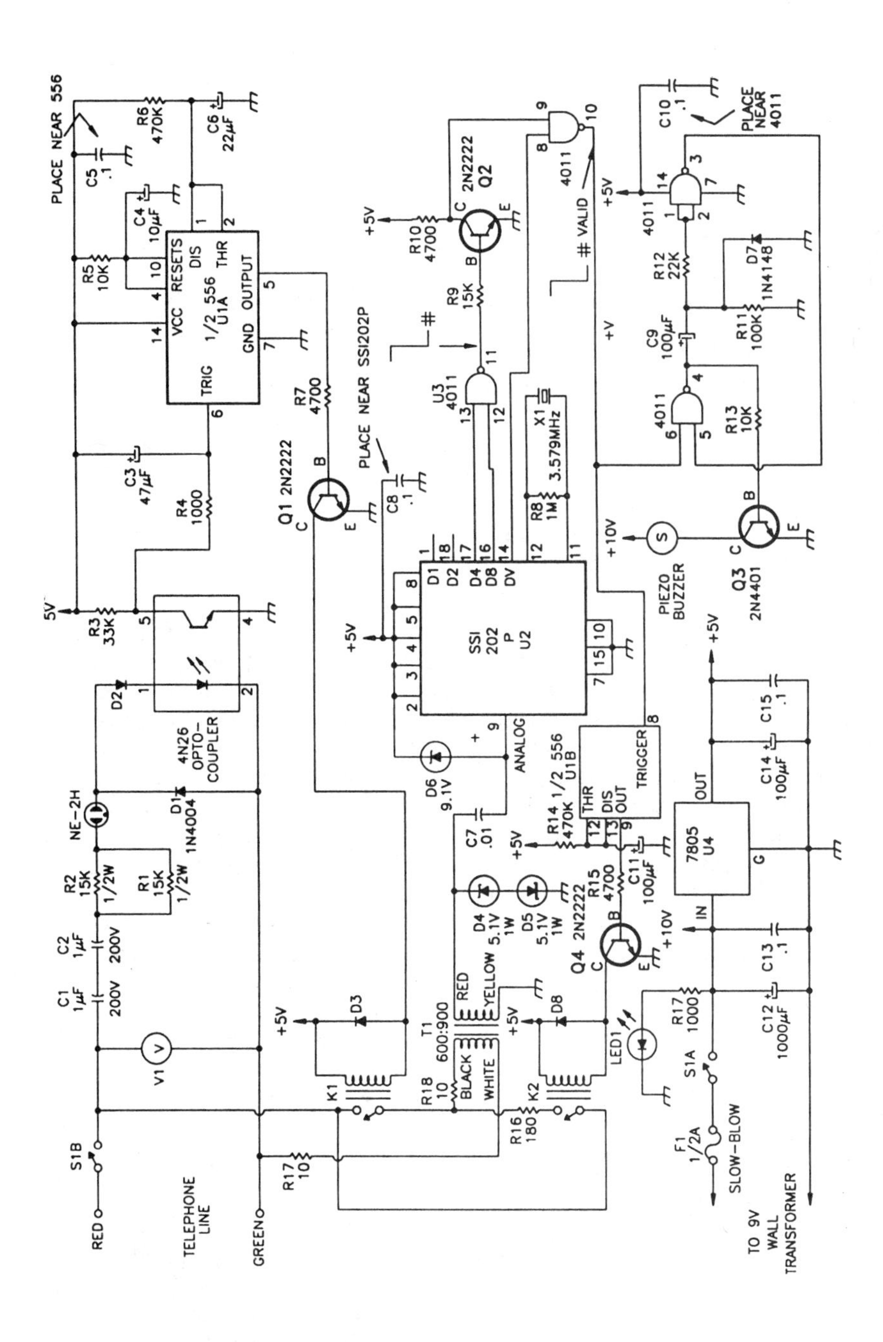

Figure 2 Complete schematic of the decoder system.

The ring detector circuit is composed of C1, C2, R1, R2, R3, D1, D2, the opto--coupler (OC1), and the NE-2H neon bulb. In normal operation, with the telephone on-hook, there will be about 50 v DC across the red and green wires of the telephone line. This voltage will generally drop to about 6-10 v when the telephone is answered. While waiting for a call, this DC voltage is blocked by capacitors C1 and C2. Two of them are used in series to obtain a greater voltage rating (400 v). They keep the ring detector from being triggered by the DC voltage on the line; however, there are many AC signals present on a telephone line, not just the ring signal. Pushing a key on the phone to dial a number generates an AC signal; talking on the phone generates an AC signal. We need to be able to distinguish between these various signals and the particular one we are after. The easiest solution is to insert a device with a high break-over threshold, such as a neon bulb, or a zener diode with a rating of 24 v or higher. All of the other AC signals on a telephone line involve relatively low voltages, generally under 6 v. The ring signal, however, is most often in the range of 80-120 v, with 90 v at a frequency of about 20 Hz being typical. By inserting the neon bulb in series with capacitors C1 and C2, we have blocked all DC voltages and all AC voltages with a peak value below about 90 v. The ring voltage is the only one remaining which can get through.

When a ring does arrive over the phone line, the AC voltage is passed through C1, C2, R1, R2, and the neon bulb to the opto-coupler. On the positive half-cycle, diode D2 passes the signal on through the coupler, which lights the internal LED. The collector of the transistor inside the coupler is normally at +5 v, due to R3; however, when the LED is turned on, this phototransistor conducts, and the collector voltage drops to a value very close to ground, indicating that a ring has been received. The purpose of diode D1 is to conduct the current around the optocoupler on the negative half-cycle. Diode D2 by itself can block the current from attempting to flow through the coupler in the wrong direction, but we need the circuit to conduct through C1 and C2 on both half-cycles, or else it would quickly charge up to the peak voltage and stop conducting. The end result would be that the circuit would conduct on only the first couple of cycles of the ring signal, and then cease to function. Diode D1 prevents this by allowing the negative cycle to also flow, discharging the capacitors. R1 and R2 limit the current through the coupler to a safe level. We use two 15K resistors in parallel to get the desired resistance and power rating.

The purpose of varistor V1 is to protect the circuitry from any high voltage transients coming over the telephone line. Though not essential to circuit operation, it is a good idea to include it. The standard 120 v varistor is not the best for this application, despite the fact that it is often used on telephone projects. A 150 v unit, such as the NTE 524V15, is a better choice; 250 v units, such as the V250LA20A, are often used because they are more readily available. The problem is that the ring voltage, which is often as high as 120 v to begin with, is superimposed on a 50 v DC level; therefore, one of the peaks of the ring voltage may cause a 120 v varistor to conduct. To be on the safe side, it is really better to use a higher valued unit.

Switch S1 is a double pole switch which cuts off both power to the circuit and disconnects the telephone line from the ring detector input. Although this last function is not really necessary (it will not harm anything if a ring is applied with the power off), there is no reason to leave the unit connected to the phone line when it is not being used; therefore, one set of the switch contacts (marked 'S1A') turns the power on and off, while the other one breaks the connection to the telephone line.

When a ring signal does come over the phone line, as stated before, the voltage on pin 5 of the optocoupler drops from +5 v to nearly 0 v. Capacitor C3 then starts to charge through R4. About halfway through the first ring, the voltage on pin 6 of the 556 IC will drop sufficiently to trigger the timer. When it does trigger, the output will go high, turning on transistor Q1, which in turn pulls in relay K1. Although the 556 output has plenty of drive current to energize the relay, the output voltage is often 1-1.5 v less than the power supply voltage. On a 5 v circuit, this is a substantial portion of the total voltage; only 3.5 v or so will go to the relay. To be on the safe side, we use the 556 to drive Q1, and use Q1 to actually switch on the relay. When the relay turns on, it connects the phone line to the isolation transformer, T1. This transformer has an AC impedance of 600 ohms, and a DC resistance of about 50 ohms. This resistance is low enough to produce an off-hook status, stopping the ring voltage and connecting the telephone line to the tone decoder through T1. The time delay of this half of the 556 is set by R6 and C6. With the values shown, the timer will hold K1 energized for about 11 seconds, and then disconnect the line. R5 and C4 provide a reset pulse on power-up.

Before the 11 seconds has timed out, a valid tone must be received by the tone decoder, or else the device will hang up. The input of the tone decoder, pin 9 of the SSI202P IC, is coupled to the secondary of transformer T1 through capacitor C7. Diodes D4 and D5 limit the voltage coming over the line to a safe level. We used 5.1 v zeners because they are more easily obtained, but 3.9 or 4 v components are better if you have access to them. Diode D6 provides added protection to the tone decoder chip, and clips any signal which would exceed the IC's input rating. This diode can be eliminated if you use 4 v zeners for D4 and D5.

The SSI202P Tone Decoder is a very specialized and sophisticated IC. Its internal operation is described in detail in the data sheet which comes with it. From the standpoint of operation within our circuit, there are five output lines with which we are concerned. Outputs D1, D2, D4, and D8 are data lines which output a binary code corresponding to the numerical value of the key pressed when a tone is received. In other words, if key number 1 is pressed on a touch-tone phone, these lines output the binary code for the digit '1.' The other output line we are concerned with is the 'DV' line ('Data Valid'). The data lines respond very quickly to a tone input, but the DV line is not raised immediately. It is delayed momentarily to verify that the signal is in fact genuine; therefore, it adds to the reliability of the circuit when this line is also used.

Since we will not be using all twelve keys, we decided to find a key whose binary code would not require exhaustive decoding. The easiest one turned out to be the '#' key. This button is assigned the binary code for the number '12.' The binary code for this key is 1100, with the '1' on the left corresponding to D8, and the '0' on the right corresponding to D1. This is the only key on a twelve button keypad which will cause both lines D4 and D8 to go high (see figure 3); therefore, since we only need one key for our application, we will only need to decode two of the data lines, D4 and D8, if we use the '#' key. The status of the D1 and D2 lines can be ignored, since they have to be low when D4 and D8 are both high, as figure 3 shows.

Key Pressed	**Output Code**			
	D8	**D4**	**D2**	**D1**
1	0	0	0	1
2	0	0	1	0
3	0	0	1	1
4	0	1	0	0
5	0	1	0	1
6	0	1	1	0
7	0	1	1	1
8	1	0	0	0
9	1	0	0	1
0	1	0	1	0
*	1	0	1	1
#	1	1	0	0

Figure 3 Output code for the SSI202P.

When a valid tone is received from the '#' button having been pressed, lines D4, D8, and DV will all go high. We can logically AND these three outputs to get a '# valid' signal, which will be used to trigger the alarm and timer, or output latch, for remote control applications. Although 3-input AND gates are available, the 2-input NAND gate is much more popular and easier to locate. The 4011 is a quad 2-input NAND gate, which means that it contains four NAND gates in each package, each with two inputs. We use one of the gates to AND lines D4 and D8. Since a NAND gate inverts the result of the AND, we must invert the signal again before 'ANDing' the result with the DV line. We could use an extra NAND gate as an inverter by tying both inputs together, but we will be using all four gates in this circuit. We, therefore, make a simple inverter out of Q2, R9, and R10. The output of this inverter will go high only when D4 and D8 are high. We then AND this with the DV line using another NAND gate. Since this gate also inverts the result, we end up with a signal on pin 10 of the 4011 IC that is always high except when a valid tone from the '#' key has been detected.

In the circuit of figure 2, two things happen when a valid '#' tone is received. One thing that happens is that the other half of the 556 timer is triggered. This starts another time delay, lasting about 50 seconds, and pulls in relay K2. This relay will connect resistor R16 and the transformer in series across the phone line. Basically, what this does is hold the line for an additional 50 seconds, so that there is time to get to the phone before it hangs up. The resistor has been inserted to keep the low DC resistance of the transformer from pulling the voltage across the phone down too low, which would significantly reduce the volume.

The other thing that occurs when pin 10 of the 4011 goes low is the triggering of a one-shot with a 7 second delay, which turns on the buzzer or alarm. This one-shot circuit is made up out of the two remaining NAND gates. To understand how this works, consider that pins 1 and 2 of the 4011 will be held at ground potential by R11 in the standby mode. This means that pin 3 will be high, since the NAND gate inverts the result. This high level is tied back to pin 5, holding it high also. Since pin 6 is connected to pin 10, which is always high unless a valid tone is present, both inputs to this gate will normally be high. This means that in the standby mode, pin 4 of the 4011 will be low. Transistor Q3 will remain off, keeping the alarm silent. When a valid '#' tone is received, pin 6 goes low. This forces pin 4 high, which turns on Q3 and the alarm; however, the transition on pin 4 from low to high causes C9 to start charging, drawing a current flow through R11. The voltage drop across R11 causes pins 1 and 2 of the attached gate to go high, which in turn drives pins 3 and 5 low; thus, even when the tone is gone and pin 6 returns high, the low on pin 5 will keep pin 4 high, until C9 has charged sufficiently so that the voltage drop across R11 drops below 1/2 Vcc. At this point, pins 3 and 5 will once again go high. Pin 4 will then go back to its low standby level, turning off Q3 and the alarm in the process; D7 and R12 protect the inputs on pins 1 and 2 from this negative transition, which would momentarily pull them below ground potential. The end result is that the presence of a valid '#' tone will not only hold the telephone line for an additional time, but also sound a buzzer or alarm for 7 seconds. Another 555 timer could have been used for this function, but we used the two NAND gates because they were already present and available.

Since this project (in our application) is not switching a 120 v load, it is not necessary to bring the line voltage into the box. We used a wall transformer to supply 9 vdc to the circuit input. This made the assembly simpler and less crowded. It also enhanced the safety of the unit, since there is no line voltage present to shock someone or to come in contact with the phone line. The power input passes through fuse F1 and switch S1 to capacitor C12, which provides on-board power supply filtering. LED1 serves as a pilot light to indicate the presence of power, while R17 limits the current through LED1. The 7805 regulator IC outputs a steady +5 v to the remainder of the circuitry. Capacitor C14 helps improve the transient response of the regulator, while C13, C15, C5, C8, and C10 all provide RF bypassing. These last three capacitors should be placed across the supply lines as close as possible to the 556, SSI202P, and 4011 ICs, as indicated on the schematic. This reduces the possibility of false triggering due to local transients.

If this device is going to be used for remote control, some changes to the circuit will be necessary. Figure 4 shows one way of doing this. Here we have removed the alarm, and disconnected the second timer (U1B) from the phone line, freeing it up to be used for controlling another device. When the device is enabled and a call comes in, the caller has 10 seconds to press the '#' key. If the tone is received, the timer (U1B) will turn on relay K2 for the delay set by the timing components Rt and Ct. When the delay has expired, K2 will de-energize. This type of circuit is handy when an automatic cut-off is desirable. Examples include coffee makers and outdoor lights.

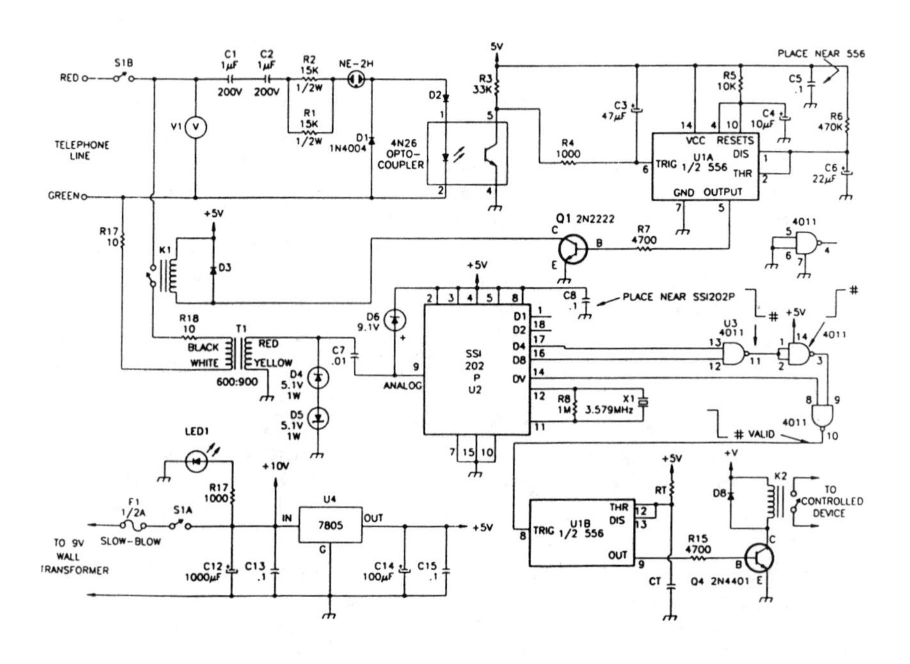

Figure 4 Circuit modified for remote control (timer output).

For simple on/off control of devices, the circuit in figure 5 can be used. This diagram shows the output changes necessary to achieve this (the remainder of the circuit is identical). First, we eliminate the second half of the 556 IC. A standard 555 timer can, therefore, be used in place of timer U1A. We then add a 4001 quad NOR gate, forming a simple Set-Reset flip-flop out of two of the gates. When the unit is enabled and a call arrives, the caller has 10 seconds to press a key. Pressing the '#' key will activate the relay, turning on the controlled device. Pressing either the '*' or '0' key will turn the relay off. The RC combination on pin 1 of the 4011 guarantees that the relay will be turned off on power-up. A push button switch can be placed across the 10 ufd capacitor to also allow a manual reset, if desired.

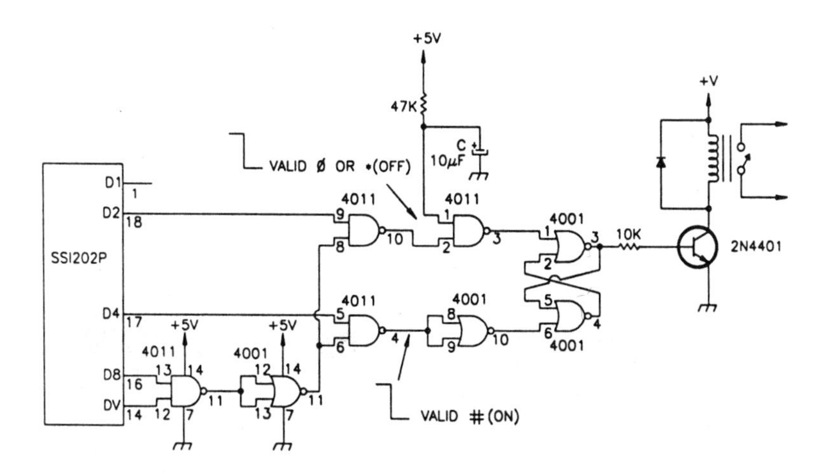

Figure 5 Two-channel remote control with independent on and off functions.

One disadvantage of using this system for remote control is that it provides no indication that the controlled device has been turned on. Although rare, you may occasionally encounter a phone where the tones are too distorted for the decoder to pick them out. Although this may not be of serious consequence when turning on a coffee maker, it could make you very unpopular if this project is used to turn the heat up in a church building on Saturday night, and you arrive the following morning to find the temperature at 42°F. One method to overcome this problem is to add a 556 IC to the circuit. The first half of this chip is used as a one-shot timer, with the timing components selected for a 2-3 second delay. The other timer section of the 556 is connected as an oscillator. The timing components should be selected for a relatively low oscillator frequency (500-1000 Hz is fine), since the telephone network has rather poor frequency response. The reset terminal of the oscillator section is connected to the output of the one-shot, and so is the coil of a changeover relay, which, when energized, will disconnect the red lead of the isolation transformer from the SSI202P, and connect it to the output of the oscillator through a .1 ufd capacitor. The trigger terminal of the one-shot would be connected to pin 10 of the 4011 in the circuit of figure 4, and to pin 11 of the 4011 in the circuit of figure 5. With either circuit, the one-shot will be triggered when a valid tone is received. This will cause its output to go high for 2-3 seconds. This will, in turn, raise the reset terminal of the oscillator, allowing it to run, while simultaneously connecting the oscillator output to the isolation transformer; thus, when the decoder receives a valid tone, it will respond by putting a 2-3 second audible tone on the line, acknowledging receipt of the command.

Many other refinements can be added when using this project for remote control. Multi-digit security codes, control of several devices, and the ability to select the number of rings are all possible enhancements; however, doing this using a 'bits and pieces' approach will quickly turn this circuit into a nightmare. It makes far more sense to use a microprocessor when adding functions this complicated.

CONSTRUCTION DETAILS

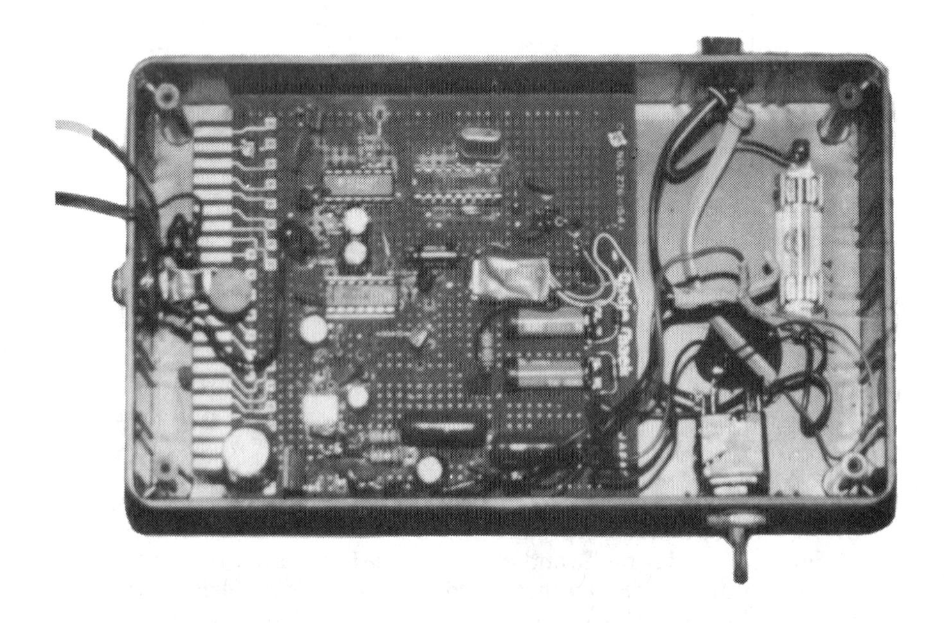

Interior view of a completed telephone remote control project.

Although this circuit is somewhat more complex than most of the projects in this book, construction is relatively straightforward. About the only thing that is critical is the placement of the 3.579 MHz crystal, X1. This component, along with R8, should be placed as close to pins 11 and 12 of the SSI202P IC as possible. Also, try to keep capacitors C5, C8, and C10 physically close to the ICs they are associated with. These prevent any transients or noise on the supply lines from upsetting the ICs. This is especially important with flip-flops and one-shots, since a transient may trigger the device, latching the error.

For this project, we somewhat simplified the power supply and construction by using an AC power adapter (wall-mounted transformer). This eliminates the need for a power transformer and bridge rectifier internal to the case. We used a Radio Shack unit (#273-1651B), which has an output of 9 v at 500 mA. At the low current level we are drawing, the actual voltage will be about 10 v. To use this power adapter, simply cut the connector off the end, and wire it directly into the circuit. On our unit, the wire with the white stripe was the positive lead.

If you decide to use this for remote control, and the device you will be controlling operates off of 120 v, you may want to go with an internally mounted transformer and bridge, since you will need the 120 v line inside the project case anyway. Another option is to use a remotely switched plug. Most hardware stores will have these. They are generally used for turning on lamps and other devices from your easy chair. They consist of a molded plug, which goes in the outlet, and a length of wire terminating in a switch. The controlled device is then plugged into the molded receptacle. For our application, you can simply cut the switch off, and connect the wires directly across the contacts of the K2 relay. For switching most 120 v loads, you will want to use a heavier relay for K2 than the one in the parts list. You can substitute another 5 v relay, or use a 9 v relay connected to the unregulated side of the supply. Some 12 v relays will also operate reliably on the 10 v available at this point.

The SSI202P IC is a highly specialized chip, and is most easily obtained by mail-order. Some possible sources are given in the parts list. The optocoupler can be obtained either through mail-order, or through special order at Radio Shack; they do not carry this part in stock, but will order it for you.

The project can be interfaced to the phone line by cutting one end off of a phone extension cord, and wiring it directly into the circuit. The cord can then be plugged into any available phone jack. If no phone jack is free in a convenient location, a 'Y' adapter can be used.

As with all projects which interface to the telephone line, FCC rules must be observed. Although this project was designed to meet the requirements for a proper telephone interface, it is technically a violation of the law to connect it directly to the line without FCC-type approval. Because this approval is such a long and expensive process, the easiest way to maintain compliance with the rules is to use an approved 'coupler.' This is a pre-approved interface that goes between your project and the line. You may want to contact your local phone company for details on whether they will require such a device, and whether they can supply you with one. If they will not or cannot get you a coupler, but demand that you have one for a project such as this, a good supplier of these items is Circuitwerkes, 1705 N. Queensbury St., Mesa, Arizona 85201.

If you decide to order a coupler, there are a couple of different types. One type of coupler is basically just a sealed unit that you plug into the telephone jack, and then plug your circuit into. The installation is fast and simple. Unfortunately, the cost of these units is often quite high, possibly costing three or four times more than the project itself. One reason for the higher cost is that the ring detector must attach to the phone line at a point prior to the isolation transformer. To make a coupler which will operate 'straight-through' while maintaining total isolation between the line and circuit, the ring signal has to be detected by the coupler, and then reproduced for the attached circuit. This added complexity adds up to a higher cost.

The other type of coupler requires a little more work, but usually turns out to be more practical for the home experimenter. This coupler generally consists of a small circuit board, to which you wire your circuit connections directly. Not only do these units cost considerably less, but their price is also partially offset by the fact that the interface will already have some of the needed circuit components mounted on-board. Usually, the isolation transformer, T1, will already be included, along with the K1 relay and diodes D4 and D5. Often, the entire ring detector circuit will be included, since this signal must be tapped off before the transformer. In this case, C1, C2, R1, R2, D1, D2, OC1, V1 and the NE-2H neon bulb can also be removed. The output of the interface's ring detector will connect directly to the junction of R3 and R4. The on-board relay will usually require either 5 or 12 v. If the relay is a 12 v model, the unregulated supply voltage of this project will usually be sufficient to operate it reliably, but a 12 v DC wall transformer can be substituted for the 9 v unit, if desired. Since most of these type couplers will only have one relay on-board (our circuit uses two for the call screening function), the best solution if you are building the circuit of figure 2 is probably to eliminate K2, D8, and R16, and tie the collectors of Q1 and Q4 together, so that both timers operate the same relay. If the volume over the phone is noticeably low until the 50 second delay expires, you may want to get around this problem by either shortening the delay (reducing R14), or by including a push button switch to ground pins 4 and 10 of the 556. If you press this switch after picking up the phone, it will immediately disconnect the relay. Do not attempt to make any changes to the interface itself, because this will void its FCC- type approval.

Circuitwerkes's model MPC-2 is one example of this type of coupler, but if you have a choice, select one with zener diodes across the transformer output, rather than standard diodes, to allow as much of the signal through as possible. This will result in more reliable tone decoding.

PARTS LIST

(For circuit in figure 2)

Semiconductors:

D1, D2, D3, D8- 1N4004 silicon rectifier (Radio Shack #276-1103)
D4, D5- 1N4733 zener diode-5.1 v-1 watt (Radio Shack #276-565)
D6- 1N4739 zener diode-9.1 v-1 w (Radio Shack #276-562)
D7- 1N4148 switching diode (Radio Shack #276-1122)
LED1- standard light emitting diode
OC1- optocoupler-4N26, 4N35, or ECG3040 (see list below for mail-order sources)
Q1, Q2, Q4- 2N2222 NPN silicon transistor (Radio Shack #276-2009)
Q3- 2N4401 NPN silicon transistor (Radio Shack #276-2058)
U1- 556 dual timer IC (Radio Shack #276-1728)
U2- SSI202P tone decoder IC (see list below for possible sources)
U3- 4011 quad NAND gate (Radio Shack #276-2411)
U4- 7805-5 v regulator IC (Radio Shack #276-1770)

Capacitors:

C1, C2- 1.0 ufd-200 v
C3- 47 ufd-16 v electrolytic
C4- 10 ufd-16 v electrolytic
C5, C8, C10, C13, C15- .1 ufd-50 v disc
C6- 22 ufd-16 v electrolytic
C7- .01 ufd-50v disc
C9, C11, C14- 100 ufd-16 v electrolytic
C12- 1000 ufd-25 v electrolytic

Resistors: (All resistors 1/4 w unless stated otherwise)

R1, R2- 15K-1/2 w
R3- 33K
R4, R17- 1K
R5, R13- 10K
R6, R14- 470K
R7, R10, R15- 4.7K
R8- 1 meg
R9- 15K
R11- 100K
R12- 22K
R16- 180 ohm-1/2 w
R17, R18- 10 ohm-1/2 w

Miscellaneous:

F1- 1/2 amp slow-blow fuse and fuseholder
K1, K2- 5 v reed relay (Radio Shack #275-232)
NE-2H- neon bulb (Radio Shack #272-1102)
S1- DPST toggle switch
T1- 600:900 ohm isolation transformer (Radio Shack #273-1374)
T2- 9 v DC, 500 mA AC power adapter (wall transformer-Radio Shack #273-1651B)
V1- 150 or 250 v varistor-GE V25OLA2OA, MOV-250-20, or NTE524V15.
X1- 3.579 MHz crystal (Radio Shack #272-1310)
Circuit board (we used a Radio Shack #276-154A)
Piezo buzzer (Radio Shack #273-066)
Project case (we used a Radio Shack #270-224)
Circuit board standoffs, optional IC sockets, misc. hardware

The mail-order companies listed below offer both the SSI202P DTMF tone decoder IC and optoisolators (at the time of this writing):

B.G. Micro
P.O. Box 280298
Dallas, TX 75228
(214) 271-5546

Debco Electronics, Inc.
4025 Edwards Road
Cincinnati, OH 45209
1-800-423-4499

JDR Micro-Devices
2233 Samaritan
San Jose, CA 95124
1-800-538-5000

Be sure to ask for a data sheet for the SSI202P IC, if available. (This part may also be listed as a 75T202 or SSI75T202P).

BRINGING UP THE UNIT

After inspecting the device for proper wiring, plug the transformer in, and connect the project to the telephone line. To test the unit, have someone call while the device is on. The device should answer during the first ring, and hold the line for about 10-12 seconds. The person calling should keep in mind that the ring they hear over the phone is not necessarily synchronized with the actual ring. The caller can tell when the device is answering by the 'click' it will make, followed by dead silence. After the device has hung up, repeat the test, only this time have the caller press the '#' key after it has answered. The piezo buzzer should go off for about

7 seconds, and K2 should close and hold the line for an additional 50 seconds or so (assuming you have built the circuit in figure 2). These reed relays are very quiet, so you will have to listen very closely to verify their operation, or a voltage measurement will confirm the same.

If the unit does not seem to function properly, check the following (all troubleshooting is based on the circuit in figure 2):

- First, verify that the power supply output voltages are correct. There should be about +10 v on the input of the regulator, and very close to +5 v on the output. If the voltage coming out of the regulator is low, make sure that Vcc and ground have not been shorted by a solder bridge. If there is no voltage on the regulator input, make sure that fuse F1 has not blown. Verify that the AC adapter is, in fact, putting out the proper voltage. If the voltage going into the regulator is correct, but the output voltage is wrong (and there are no shorts on the output), replace the regulator IC. If the voltages seem correct, also make certain that the power supply voltage is present at each of the ICs.
- If the power supply voltages seem correct, verify operation of the alarm and one-shot circuits. Using the circuit in figure 2, momentarily short pin 10 of the 4011 IC to ground. This should cause the buzzer to sound for about 7 seconds, and cause relay K2 to energize for about 50 seconds. If the buzzer does not operate properly, carefully check all connections on pins 1-7 of the 4011 IC. If pin 4 of the 4011 goes high when pin 10 is grounded, but the buzzer does not go off, check both Q3 and the buzzer itself. Also make sure that diode D7 is not installed backwards. If no problem can be found, replace the IC. If the 50 second delay does not operate properly, check the connections on pins 8 through 13 of the 556 IC. Make certain that pin 10 is near +5 v. If pin 9 goes high when the timer is triggered, but the relay does not energize, check Q4, and make certain D8 has not been installed backwards.
- To verify the operation of the other timer, ground pin 5 of the optocoupler for at least 1/2 second. Relay K1 should energize, and then turn off after a 10-12 second delay. If not, carefully check pins 1-7 of the 556 IC, and make certain that pin 4 of the 556 is near +5 v, and pin 6 goes to ground when pin 5 of the optoisolator is grounded. If pin 5 of the 556 goes high, but K1 does not energize, check Q1, and make sure D3 is not installed backwards. If no other problems can be found, replace the 556.
- Once operation of the alarm and timer stages has been verified, we can test the tone decoder circuit. The easiest way to do this is to momentarily ground pin 10 of the 4011, so that the buzzer goes off and K2 turns on. After the buzzer shuts off, pick up the phone and press the '#' key. If the tone decoder is working correctly, the alarm will go off again. If it does not, carefully check all wiring around the SSI202P and pins 8-13 of the 4011. Make sure that the crystal and R8 are connected correctly to the decoder, and make sure that diodes D4, D5, and D6 are

installed in the right direction. Carefully check all connections between T1, K1, and the phone line. If the DV, D8, and D4 lines do not go high when the tone is present, try another decoder chip. If these lines all go high, but pin 10 of the 4011 does not go low, check Q2. If the voltages on the inputs to the 4011 gates are correct, but the output does not switch, replace the 4011.

- Lastly, we must test the operation of the ring detector. In the circuit of figure 2, verify that the neon bulb flashes when the phone rings. If it does not, you have an open circuit somewhere between S1B and pin 2 of the optoisolator. Make certain that pin 2 of OC1 has been connected to the phone line input. If the wiring of the detector seems correct, the most likely cause of failure is the neon bulb itself. You can test C1 and C2 by shorting one of them out and applying a ring. If shorting one of them makes the circuit work, the one that is shorted is bad. If the neon bulb starts to fire, but does not flash for the entire ring, either D1, D2, or OC1 is probably at fault. If you live in an area where the ring signal is abnormally low, the neon bulb may not be able to fire. In this case, replace the bulb with two back-to-back 24 v zener diodes.
- If the neon bulb flashes as it should, but the 10-12 second timer does not trigger, make sure pin 4 of OC1 is grounded. Then measure pin 5 of OC1 as a ring is applied. If the voltage does not go low, replace the optocoupler. If the voltage on pin 5 still does not go low, the ring voltage in your area may be less than normal. If necessary, place an additional 15K resistor in parallel with R1 and R2, or substitute the zener diodes as mentioned above. If it does go low, but pin 6 of the 556 does not go low enough to trigger 1/3 Vcc before the ring ends, replace C3 with a smaller capacitor, such as a 22 ufd unit. The smaller the value of this capacitor, the shorter the ring will be before the unit answers.
- If you are using a coupler such as the MPC-2, which includes the ring detector, the 'ring' terminal should go low when a ring arrives. If it does not, the interface unit is probably bad, or you may have an oddball coupler, which may require a slight modification to this circuit for proper operation.

☛ Dear Reader: *We'd like your views on the books we publish.*

PROMPT® Publications, an imprint of Howard W. Sams & Company, is dedicated to bringing you timely and authoritative documentation and information you can use.

You can help us in our continuing effort to meet your information needs. Please take a few moments to answer the questions below. Your answers will help us serve you better in the future.

1. What is the title of the book you purchased? ______________________
2. Where do you usually buy books? ______________________
3. Where did you buy this book? ______________________
4. Was the information useful? ______________________
5. What did you like most about the book? ______________________
6. What did you like least? ______________________
7. Is there any other information you'd like included? ______________________
8. In what subject areas would you like us to publish more books?

   (Please check the boxes next to your fields of interest.)

   - ❑ Amateur Radio
   - ❑ Antique Radio and TV
   - ❑ Audio Equipment Repair
   - ❑ Camcorder Repair
   - ❑ Computer Hardware
   - ❑ Computer Programming
   - ❑ Computer Software
   - ❑ Electronics Concepts Theory
   - ❑ Electronics Projects/Hobbies
   - ❑ Home Appliance Repair
   - ❑ TV Repair
   - ❑ VCR Repair

9. Are there other subjects not covered in the checklist that you'd like to see books about?

   ____

10. Comments ______________________

____

____

Name ______________________

Address ______________________

City ______________________ State/Zip ______________________

Occupation ______________________ Daytime Phone ______________________

Thanks for helping us make our books better for all of our readers. Please drop this postage-paid card in the nearest mailbox.

For more information about PROMPT® Publications,

see your authorized Sams PHOTOFACT® distributor.

Or call 1-800-428-7267 for the name of your nearest PROMPT® Publications distributor.

Imprint of Howard W. Sams & Company

2647 Waterfront Parkway East Drive,

Indianapolis, IN 46214-2041

BUSINESS REPLY MAIL

FIRST CLASS MAIL PERMIT NO. 1317 INDIANAPOLIS IN

POSTAGE WILL BE PAID BY ADDRESSEE

HOWARD W. SAMS & COMPANY

2647 WATERFRONT PKY EAST DR
INDIANAPOLIS IN 46209-1418